THE SECRETS
OF METALS

THE

SECRETS OF METALS

By

WILHELM PELIKAN

Translated from the German

By

CHARLOTTE LEBENSART

ANTHROPOSOPHIC PRESS
Hudson, New York

Originally published in 1952, this book first appeared
with the title, *Sieben Metalle* in the second edition of
1959. A translation of this edition is published here with
the kind permission of the Philosophisch-Anthroposophi-
scher Verlag, Dornach, Switzerland.

Printed in the United States of America

PREFACE TO THE
FIRST GERMAN EDITION

This work is an attempt to throw light upon an important segment of the world of matter by placing it in a large perspective. The world of metals, despite all the achievements of chemistry and physics, is still full of mystery. The history of their discovery, extraction, and utilization is a significant chapter in human development. With the help of metals man has established his position in the world and has transformed the face of the earth. The concord of his bodily functions depends, in many important processes, on marvellous metal-borne effects. Every decade of research acquaints us with new facets of a cosmos of metals within us. In the world outside we come again and again upon new deposits in the earth, which enable us to advance in civilization; in the inner world of the body, ever new layers of activities permeated by metals are lifted into our consciousness. We not only breathe with iron, but we need copper to form our blood and cobalt to escape pernicious anemia. As the methods of investigation become more refined, we constantly discover more metals to be regular components of our bodies. We find them, however, not as building blocks in the grosser sense, but as instruments by which our human entity carries out significant physiological activities.

A consideration of the metals, therefore, may count upon general interest. The engineer and the artist need to know the nature of these materials, in which they must incorporate their works. The physician wishes to find a deeper connection between natural substances and the body so that he may arrive, in a rational way, at remedies wherein the forces of the universe are brought into a harmonious and helpful accord with suffering man. Lastly, the educator will be able to awaken strength and courage in

v

the growing youth if he can tell him, "What is of service to you in your body, the body that enables you to assert your soul and spirit in the physical world, rests as yet untouched in the depths of the earth. It waits upon your creative will, meaningfully embedded within the whole of earth life. What you need in order to breathe, to form your blood, has its tasks also within the whole of outer nature. You are the one who releases it, and with it you create the works of man. Reintegrate it again within the world-whole."

PREFACE TO THE
SECOND GERMAN EDITION

Seven years after the appearance of the first edition (long since out of print), this second edition is presented to sympathetic readers. Advancing research has in the meantime found new and interesting details, but these confirm rather than change the pictures of the Seven Metals selected in the first edition. More refined methods of investigation have now established the presence in the organic world of almost all metals, in various degrees of dilution down to mere traces. For example, the manual of Bertalanffy lists the following metal concentrations in human plasma:

magnesium, iron, aluminum, zinc .... D 1 to D 2
copper, manganese .... D 2 to D 3
arsenic, titanium, vanadium, chrome, nickel ....
strontium, lithium .... D 3 to D 4
silver, cobalt, rubidium, tin, molybdenum .... D 4 to D 5
gold, uranium .... D 5 to D 6
uranium in bones .... D 6
uranium in muscles and teeth .... D 9
radium .... D 11

D represents the decimal potency used in homeopathy, and is used here because it so clearly expresses the relationship in figures. Thus D 1 corresponds to a dilution of 1 to 10, D 2 of 1 to 100, etc. In the above-mentioned manual the concentrations are expressed in powers of ten: instead of D 1 the expression 10^{-1} is used; instead of DX, 10^{-x}, etc.

Compared to the quantities ascertainable in the inorganic world, the above figures represent powerful condensation processes; one must say that the life processes seize and suck in the world of metals in a vigorous way. In the plasma of ten sea animals, for example, an average increase in the antimony content by 300, and in the vanadium content by as much as 28,000 has been established.

An unchanged reprinting of the first edition dealing with the classical Seven Metals seemed clearly justified. But the author and many readers felt the need for an elaboration of this work. Therefore, six new chapters have been added to the original, entering untrodden fields. The material on heretofore unknown "radiation effects of metals" is also new. This material is based on a reference by Rudolf Steiner, who used these effects therapeutically.

A high sense of responsibility arises from such an approach. If the present work contributes only modestly to this goal, it may at least serve as a stimulus to a more perfect undertaking.

The method pursued in the following chapters is a phenomenological one, in which we try to look upon the natural object in the Goetheanistic sense. But the light thus falling upon the object flows from the work of Rudolf Steiner, from the anthroposophical science of the spirit. Of its well-nigh boundless wealth a small part only could be used. The reader therefore is expressly directed toward the whole of this work, particularly the fundamental books. Working with these, one will become aware of the new language that the phenomena of nature acquire.

CONTENTS

THE SECRETS
OF METALS

I

THE METALLIC CONDITION
ON EARTH

Among the great variety of phenomena on the earth, the metallic nature stands out as something extraordinary. The metals are lustrous, polished, durable, flexible yet solid, and impress us with their weight; we feel they have significance and dignity. Not only are we aware of them by our usual sense perception, but our "unknown" senses (to use Goethe's expression) are also aware of them. Their obvious properties have caused metals to be made into tools in the service of mankind, and have even led to the naming of the Copper Age, the Bronze Age, the Iron Age. What the "unknown" senses experienced in the past revealed to man a deep inner relationship between himself and the metals that led to their use as charms and ornaments, and as vessels for the celebration of various cults. Metals, however, have their relationships not only with man but also with all forms of earth existence. As we compare these relationships we become aware of the great transformations that the metals undergo as they function in the various kingdoms of nature.

In the mineral kingdom we find metals everywhere. They are indispensable constituents of many minerals and rocks. Nevertheless, the laws of the mineral world rarely allow the metallic qualities of the metals to be seen. Pure native metals—gold, silver, platinum—are but rarely found in nature, hence they have always been considered precious. More frequently,

3

we find metals in the form of ores, in combination with other elements, which only sometimes allow the metallic nature to reveal itself. The semi-precious metals have been extracted from their ores by primitive metallurgical practices for ages, and always a deposit of tin or a vein of copper ore has been worth a fortune. The common rocks and soils, however, also contain certain metals, but these are so well concealed that for ages their presence remained unsuspected. The discovery of new techniques and the application of new knowledge were necessary before man could extract aluminum from clay, calcium from chalk, magnesium from epsom salts, and sodium from rock salt.

Searching with the refined methods of the modern chemist, we really find all the metals present everywhere; most of them, however, are subtly distributed. The rocks contain them, though in various concentrations and mixtures. The soils derived from rock contain them, above all the ordinary plow-land. The waters of the oceans are tinged with them, each in a different way. In still finer dilutions they must surely be present in the air. During an expedition to Greenland, Wegener noted that the freshly fallen snow lost its pure whiteness after a short time because of a thin layer of dust, which contained iron and was doubtless the meteoric dust that constantly sinks down from the cosmos into the atmosphere. In a form no longer physical, the light streaming to us from the universe has been recognised by spectroscopic analysis as "metal light" of the most diverse kind. Indeed, the sunlight shows the spectroscopic lines of iron, gold, and most of the known metals. We might term it "metallic essence in light form."

If in imagination we approach the earth from the cosmos, we come upon places of increasing concentration of the metallic nature until at last it confronts us in the compact "physicalized" form of the ore deposits in the depths of the earth. Everywhere the metals are present. But just as the body shows the presence of blood wherever we cut it, yet in greatest abundance in the veins and finally in the heart, so does the earth contain gold, silver, copper, iron, etc. everywhere, but in few places do these appear in significant quantities. We discover, too, that each metal has its own peculiar distribution over the earth. If we mark on a globe the various localities containing deposits of gold, copper, uranium, etc., a different pattern arises for each metal. Something in the nature of an organic structure becomes visible. Behind the existing lead or gold we see, as it were,

4

the lead organ or the gold organ of the earth. We shall later, in describing the individual metals, enter into details and become more specific.

But not only the mineral kingdom is permeated by metals. Plant, animal, and man are equally so. Here we touch upon a field that will yield its greatest results only in the future, but we already know that the most varied functions of life require the metals. Without magnesium, plants could not form their chlorophyll or build up their bodies out of air and water. The lower animals could not breathe without copper, nor the higher animals and man without iron. Lack of cobalt in the soil causes serious epidemics in cattle, and in man it appears in the disease of pernicious anemia. Throughout the human organism we find gold, silver, mercury, tin, etc., in minimal traces, each metal in a different distribution, in one organ more, in another less. Some play an essential role in constructing certain organs, where their grosser materiality is needed to serve as building blocks: calcium, for example, as calcium carbonate and phosphate of lime in bone formation, and as calcium fluoride in tooth enamel. But for the most part metals are active in the body in extreme degrees of dilution, thus indicating that what counts is not their materiality but their dynamic way of working. Conversely, it is apparent that the body is "accessible" to the most varied metals, if only they are rendered into a condition so fine as to suppress their physical aspect but allow their dynamic nature to unfold. The metals, it becomes obvious, are endowed with a certain "organotropy," a profound relationship to the organs whereby a particular metal is drawn to a particular organ.

By following up this organotropy of the individual metals by modern means, we can regain an understanding of correlations between them and the inner organs that were known in ancient times and from which came many clues for their use in healing. We will see that these correlations reveal the nature of the metals just as clearly as the physical and chemical qualities heretofore considered so important. This will be attempted here, so far as possible.

The metals affect man not only in his body but also in his consciousness, in his soul and spirit. They speak a language that conveys their nature more impressively as these effects move into higher spheres of existence. What is a mere indistinct stammering in the realm of the organic becomes more articulate as it is taken up and utilized by more perfect realms—through the realm of the merely living plant, into that of the en-souled animal, and finally into that of man. For the higher a being rises and the more it is able to express its own essence, the better it can express the regions of the world from which it derives.

Within the mineral world the metals lead a paradoxical existence. They are, to be sure, important components of the earth, but the normal earth processes are far more inclined to conceal or destroy the metal condition than to produce it or even let it continue to exist. A metal that has the strength to prevail against the earth processes so that we find it in pure form, such as gold, silver, or platinum, is so unusual that we elevate it to the rank of a precious metal. Mostly, however, we are offered only the ore form, which sometimes still has at least a semblance of being metallic. In the processes of metal extraction we combine the heavy, the ponderable, the earthy with the "imponderables," the weightless energies irradiating space, and only then does the pure metal, freed from its fetters, arise from the ore. In the case of the semi-precious metals, such as copper, mercury, or tin, this is easy, needing only a minor application of these imponder-ables. A far stronger effort is required with common metals, such as iron, zinc, or antimony, and tremendously powerful applications are necessary to free the metals hidden in rocks and soils. For this reason mankind has been rather late in becoming acquainted with magnesium, aluminum, and calcium. Lastly, the alkaline metals yielded by the salt nature of the earth, such as sodium, potassium, and their relatives, are not only obtained with an enormous expenditure of energy but remain highly unstable, highly artificial, striving to revert to salts with the greatest possible speed. Be-cause of their softness, low melting point, and instability, we must call them stunted in their properties, caricatures of the real metals. They must be protected by glass walls or under petroleum, because every drop of water or breath of air threatens them. Truly a laboratory existence!

Thus the tendency of the earth processes is not to liberate the metallic quality, but to annihilate it, to burn it, to reduce it to ashes or at least to rust, to patina. With full justification we may therefore conclude that the metallic quality is, fundamentally, a stranger to the earth. It is a guest rather than a citizen. Actually, it can neither be explained nor understood from the point of view of the earthly-mineral. Where, then, is its exist-ence rooted? Has it perhaps a home other than the earthly? Does it be-

6

long to another world? Can it be that its task is to fuse the spheres of activity beyond the earth with the sphere of the earth? Should we designate the metals as earthly-cosmic, comparable perhaps to the plant, which, though tied to the earth with its roots, nevertheless belongs to the cosmic light through its leaves? Out of its seed, which, because of its weight, falls victim to the earth, to the ponderable, the plant is released to return to its true form by the imponderables that stream into it from the cosmos, such as light, warmth, etc. In this sense the ores would be seeds and the metals would be plants, freeing themselves from their seeds through interweaving with the imponderables.

The ancients spoke of three processes in the mineral kingdom: Salt, Mercury, Sulphur. Salt to them was the ponderable matter, subject to the earth forces. Sulphur was the substance saturated with the imponderables, the forces raying in from the world circumference. Mercury, however, was a substance open to the rhythmic interplay of forces from both sources, one radiating out from the earth and the other radiating in from the cosmos. (In this respect all metals are mercurial.)

In the last century the earthly aspect of the metals was thoroughly investigated. We would now like to look at the cosmic aspects of their nature in order to do better justice to the whole and to prepare our further considerations.

II

COSMIC ASPECTS
OF THE METALLIC NATURE

Our century seems to be slowly rediscovering the relationships of the cosmos to the life of the earth. The cosmic rays in physics, the connection between the moon and the life-rhythms of certain plants and animals, the changes in certain human blood reactions according to the movements of the sun, are all indications of this. We might even say that new sciences are being founded, such as cosmo-biology. The influence of planetary aspects on radio reception is watched. Technology seeks for ways to penetrate into the highest layers of the atmosphere in order to intercept the cosmic influences before they are altered by earth conditions. Travel in outer space, a personal invasion of cosmic space, is seriously contemplated. Less and less is the earthly atmosphere regarded as our boundary; it is looked on more as a covering layer having many points of contact with the cosmos, a "sheath organ" with the function of receiving and "digesting" the rays that constantly flow in from the universe.

This comparison with a living organism forces itself upon us inevitably, and has been impressively worked out in the book, *Earth and Man*, by Dr. G. Wachsmuth. Just as the embryonic sheaths or membranes transmit the forces of the maternal organism to the child, and in the process of transmission also transform them (untransformed they would be deadly), the diverse enveloping layers of the atmosphere transmit and transform the rays that nourish the earth and bring movement into all of its life. Without this transformation the rays would destroy everything on earth. Thus, the form in which we receive the gifts of the cosmos is not the form that they have in their own sphere before reaching the earth. We have chosen the simile of the connection of the maternal organism with the child because the earth is indeed born out of the cosmos. From the cosmos the earth receives not only processes that sustain its life, but also substances that build it up. The meteors that continually invade our atmosphere conclude their journey to the earth as metallic-mineral messengers from the universe.

From this we can clearly see that the metallic nature is not only a property of the earth, but that it extends out into the cosmos. We become aware of the heavenly origin of certain minerals. Meteoric iron contains iron, cobalt, nickel, copper, and several other metals in a form that cannot be found or artificially produced on earth. Though iron is the commonest metal on earth, we find it only in its ores, never in pure form (with the exception of one remarkable deposit on a small island near Greenland). As a pure unadulterated metal, iron comes to us not from the earth but from the cosmos. In its metallic form it is a stranger upon earth, a son of the cosmos.

If one makes a disc from a ribbon of magnesium, holds it in one's line of vision so that it appears just to cover the disc of the sun, and ignites it, it will give off the same intensity of light as the sun itself. But with the help of this same magnesium the sun's energy is incorporated into the chlorophyll of the leaf. Every ray of light ends, as it were, in the chloroplast as a granule of starch. But iron is also necessary for the formation of chlorophyll, that cosmic organ of the plant. Thus the plants need the metals in order to connect the cosmic aspect of their existence with the earthly.

But there are also life spheres of a cosmic nature within the earthly—cosmic enclaves, as it were. Here also the metals play a many-sided role. We find them in the bodies of beings that rank above the plant. In fact, what distinguishes these from the plant is just their ability to carry out independently those processes that the plant must carry out in conjunction with the cosmos, such as the formation of starch in the liver. The plant has no inner organs; leaf and cosmos belong together. The animal has a wealth of inner organs, an inner cosmos of organs.

The plant receives its sprouting, blossoming, and ripening from the outer universe; the animal, by way of its inner organs. It substitutes the rhythm of its organs for the cosmic rhythms. It is, therefore, not improper to consider these organic rhythms as something cosmic that has turned inward. Thus we become aware that the organ-cosmos uses the metals no longer for their material, but for their dynamic effects. Copper prepares the hemoglobin formation, iron builds up the hemoglobin as well as the breath ferments. In short, the inner cosmos is permeated by metallic activities.

* * *

If we pursue phenomena such as these, the following question arises. If to this day certain metals are showered from the cosmos onto the earth, could not the metals now resting in the womb of the earth as ores have come as a gift from the cosmos to the youthful earth in the first days of creation?

Here we touch upon an age-old insight of mankind. This did not derive from thoughts such as ours, but from a genuine "perceiving" with ancient soul forces long since extinguished. It asserted specifically that iron was a metal of Mars; copper was a Venus metal; gold was something sun-like; lead was assigned to Saturn; tin belonged to Jupiter; silver was governed by the moon; quicksilver was given the old name of Mercury. An understanding of this ancient wisdom can be regained now that Rudolf Steiner's modern spiritual-scientific research has come into existence. But let us see whether modern physical science, through its own knowledge of the metals, does not find points where it agrees with the science of the spirit.

Modern research is well acquainted with the weight relationships of the metals, the so-called "atomic weights." Now let us take the atomic weights of the seven metals that the ancients connected with the five planets, the sun, and the moon, and arrange them in a circle according to their magnitude so that we obtain a heptagon as in Figure 1. The result is a significant arrangement.

The singular harmonies that become apparent when the basic chemical substances (the elements) are grouped according to their atomic weights

10

Fig. 1

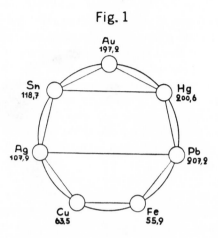

were, of course, discovered in the last century by Lothar Meyer and Mendeleyev, as the so-called "periodic system" of the elements. We proceed in the same way with the seven metals, except that we use a circle instead of the straight column of the periodic system. Although this arrangement is possible only because modern methods of analysis have established the atomic weights, a system of relationships is hidden in it that was familiar to the ancients. Thus the new arrangement combines old knowledge with modern research.

The atomic weights of the seven metals are:

METAL	LATIN	SYMBOL	ATOMIC WEIGHT
Iron	Ferrum	Fe	55.85
Copper	Cuprum	Cu	63.45
Silver	Argentum	Ag	107.88
Tin	Stannum	Sn	118.7
Gold	Aurum	Au	197.2
Mercury	Hydrargyrum	Hg	200.6
Lead	Plumbum	Pb	207.2

In Figure 1 we start at the right corner of the base line of the equilateral heptagon with iron (Fe) and its atomic weight of 55.85, and proceed left with Cu, etc. This arrangement reveals that above, in the center, gold is enthroned in a special position. The other six metals confront each other in pairs; first tin and mercury, then silver and lead, and finally copper and iron. Polarities appear that are lost in the usual arrangement of the periodic system. Nevertheless, these polarities exist in nature, and also in man.

Let us begin with the base line of our heptagon. Copper and iron, though not considered to be close neighbors in ordinary chemistry, are spoken of by nature almost in the same breath. The most important copper ore (chalcopyrite) is also, as copper-sulphur-iron, a distinct iron ore. It is as though the mineral nature wanted to direct our attention to something that achieves its full expression only in the realm of the living. For in that realm these two metals, and no other, are involved in the breathing process by which animal and man permeate their fluid organization with air. The hemoglobin used in breathing by many lower water-creatures, such as mussels, snails, crabs, and cuttlefish, is a compound of protein and copper, while the higher animals and man breathe with a similar iron compound as their hemoglobin.

Turning now to the second horizontal line of our heptagon, we find silver and lead. Ordinary chemistry does not view these two metals as polarities, but nature demands it of us. Lead sulphide (galena), the most important lead ore, always contains a small quantity of silver glance (argentite). Indeed, by virtue of this fact it is also the most important silver ore. Far and away the greatest quantities of silver produced are extracted from galena. Silver nestles in lead's lap in nature. Pure silver ores or deposits are exceedingly rare. In the human body too, silver and lead appear as polarities, but they do not show this as normal physiological elements in the rhythmical system, as iron and copper do. It comes out in their therapeutic effects. Lead increases the degenerative processes emanating from the nerve-sense system; it promotes hardening and paralysis. Silver stimulates the regenerative processes of the metabolic system, enlivening and refreshing it. (More details will be given in the description of the individual metals.) In silver, too, we can find antidotes against the poisonous effects of lead.

Tin and mercury likewise show polarities in the body. Both work in

12

the fluid organization as remedies, though in entirely different ways. More will be said about this later in the individual descriptions.

In our diagram gold quite properly stands alone. In nature, too, it stands by itself, being distributed over the earth according to its own laws. In the body it works as a remedy for the heart, the center of the rhythmic system.

Now let us draw inside our heptagon three isosceles triangles: the left one connecting the points of copper, silver, and mercury; the right one, the points of iron, lead, and tin; the center one, the points of copper, iron, and gold. Again we find correlations between the metals, not only in their natural attributes but in their relationships to human organic processes.

Fig. 2

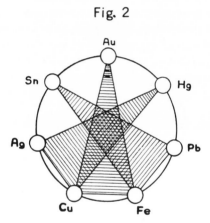

Copper, silver, and mercury, the metals of the left triangle, appear in nature together with antimony, in the so-called fahl-ores-grey copper ore or tetrahedrite, typical sulphur minerals. The three metals are semi-precious to precious; soft, pliable, plastic; excellent conductors of heat and electricity (using solid mercury for this comparison). The melting points of the copper-silver-mercury series decline from 1083° C. to 960° to minus 39° in that order.

The metals of the right triangle, iron, lead, and tin, are not precious; they are poor conductors of heat and electricity; and their melting points decline from 1530° to 327° and 231° respectively. In the earth they keep apart, seldom appearing in the same place, even as combined ores. Thus the triads of the left and right triangles are strong opposites, polarities.

The middle triangle comprises copper, iron, and gold. Gold can be found in copper ores as well as in pyrites (ferrous sulphide). As a metal, gold resembles no other metal so much as copper, which might therefore be called its plebeian half-sister. The melting point of gold (1064°) also comes unusually close to that of copper (1083°). But the most important gold salt, trichloride of gold, resembles most closely trichloride of iron. Furthermore, copper, silver, and mercury are alkaline, and hydroxide of silver forms a strong lye. Iron, lead, and tin, however, form acids whose salts are known as ferrates, stannates, plumbates. Here, too, gold stands in the center, reconciling the polarities and harmonising them.

But also in their influences on and relation to the human organization these triads show something striking. Silver, mercury, and copper work especially upon the metabolic system. They stimulate its regenerative functions, accompanying the nutritive processes as far as the blood formation, in which copper actually plays a part. Lead, tin, and iron work on the upper region, the nerve-sense organization. They are related to the degenerative, hardening, shaping processes. Iron works through the respiratory process into the rhythmical organization, constituting the breath pigment, and the triad copper-iron-gold is mirrored in the rhythmic system, gold inclining to the heart as the center, iron to the respiratory aspect, and copper in the direction of the metabolism. This threefold structure of the triangles of our heptagon thus gives a metallic reflection of the threefold organization of man.* This thrice differentiated picture of the seven metals, these relationships of the triads lead-tin-iron, silver-mercury-copper, and iron-gold-copper, reflect the upper, lower, and middle man according to their own laws.

* * *

What we have said amply demonstrates that our arrangement of the metals is "marked on their forehead," since it brings to light many things about the metals in nature and in man that otherwise remain concealed. It

---

* Rudolf Steiner, *Riddles of the Soul.*

corresponds in a certain sense to the true essence of the metals. Most significant, however, is the fact that a system of cosmic relationships is hidden within it. Something of the origin of the metals shines through.

To demonstrate this, let us once more start with the circle divided into seven parts. But now we construct a seven-pointed star by beginning at lead, omitting two intervals, and drawing a diagonal to tin, and so on, so that the line continues to iron, gold, copper, mercury silver, and back to lead. (Only two stars can thus be drawn in a heptagon: one acute and the other obtuse. Both have a significance, as we shall see.) At the same time we draw beside each of the metals the sign of the planet to which, from time immemorial, it has been assigned: Saturn by lead, Jupiter by tin, Mars by iron, Sun by gold, Venus by copper, Mercury by mercury, the Moon by silver. The lines of our star now yield the sequence: Saturn-Jupiter-Mars-Sun-Venus-Mercury-Moon.

Fig. 3

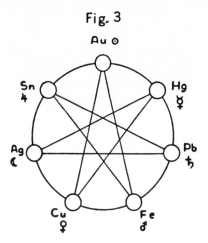

In this sequence there appears the spatial arrangement of our planetary system, as seen from the earth and as shown in the so-called Ptolemaic system. Saturn, Jupiter, and Mars are the outer planets, Venus and Mercury the lower ones, while the moon is the heavenly body closest to the earth. Whenever the inner planets stand in an inferior conjunction before the sun and the outer ones in conjunction behind the sun, we have the exact se-

15

quence shown in Figure 3, beginning with Saturn and ending with the Moon. The spatial order of the cosmos thus emerges in our metallic order. The metals belong not only to the earth, but to the entire cosmos. This is what our diagram demonstrates.

But another and entirely different chain of relationships is hidden in it. This becomes visible if, starting again from the lead-Saturn point and now omitting only one interval, we draw the diagonals that give us the second, blunter, seven-pointed star. The line now runs from the Saturn-lead point to Sun-gold, to Moon-silver, to Mars-iron, to Mercury-quicksilver, to Jupiter-tin, to Venus-copper, and back to the beginning at Saturn-lead. This sequence, Figure 4, has nothing to do with space, but shows a structure in time, a rhythm of world evolution.

Fig. 4

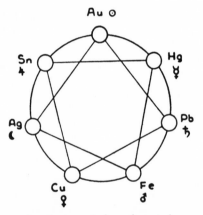

Our existence evolves in time. It depends entirely upon the rhythms resulting from the cooperation of the movements of earth and stars. The year reflects the sun rhythm, the month the rhythm of the moon. By years and months we count our lives. The succession of days brings us, each week anew, a consciousness of the rhythm of seven, containing the sun, the moon, and the five planets, as they follow each other in the blunt star in Figure 4. We start with Saturday, whose English name points clearly to Saturn. We proceed to Sunday and Monday, named after Sun and Moon in

16

English and also in other languages, since Luna is spelled out in the *lundi* of the French and the *lunedi* of the Italians. Tuesday has behind it Ziu, the Germanic god of war, who was Mars to the Romans and appears in *martedi* in Italian and *mardi* in French. (To the ancients the stars and the planets were abodes of the gods in the same way as the earth is the star of human habitation.) Then follows Wednesday, formerly Wotansday, still retaining its pure form in English. The Romans felt Wotan to be the same being as their Mercury. In the Romance languages the French have retained him in *mercredi,* and the Italians in *mercoledi.* Thursday belongs to Donar òr Thor, as is evident in the English form. The Roman thunder god is Jupiter or Jove, as shown in *jeudi* in French and *jovedi* in Italian. Freya, the Germanic goddess, corresponds to Venus of the Romans. Friday, *Freitag, vendredi, venerdi,* point to her. With the names of the days, each day of our lives, we are reminded of the cosmic rhythms.

The ancient wisdom named our units of time after the stars. The experience of every week reminds us of the evolution of the universe, of the secrets of cosmic becoming, and of the fact that the earth and all its creatures, at every moment, are in constant connection with the spatial configuration of the cosmos and the laws of evolution in time.

The days of our lives follow the rhythm of world evolution; our earthly history images the world history. Once upon a time, out of an old knowledge, the harmonious accord of the evolution of man and earth could still be felt. Every substance on earth had to partake in it. Now we try, in a new way, to grope toward the secret resting in every earthly substance as well as in man and the extra-terrestrial world. From the time of Goethe there linger some hints* that in the extra-human, in the realms and beings of nature, we can find the milestones we have left behind on our way to becoming man. In our days, for the first time, this has been expounded by Rudolf Steiner in a way suited to modern consciousness. Let us steep ourselves in the chapter on the evolution of the world in his book *An Outline of Occult Science.*** If history is to do more than register basically unconnected events, if it is to be a perception of spiritual impulses in evolution,

---

* "The world is a precipitation out of the nature of man."
"We are related to all parts of the universe, to the future as well as to the distant past."
"The world is the sum of what is past and what is present."—Novalis, *Fragments.*
** Anthroposophic Press, Spring Valley, N.Y.

its true task will be the development of a sense for the observation of the "spiritual score" that unfolds as history into time, but is itself above time and of a spiritual nature. One who can read the spiritual laws of evolution will find his way back into the world-past, even where the outer documents fail. From such reading come the accounts in the above-mentioned book. According to these accounts, our earth, together with the creatures (the three kingdoms of nature, and man, and also its entire cosmic environment) constitutes, as it were, the fourth movement of the symphony of world creation. The three preceding movements, having prepared the fourth, have faded away. Or, differently expressed, three world creations have evolved into physical forms of existence out of the spiritual and have faded back into the spiritual, as a plant unfolds from its seed and fades at the year's end, presenting to the new year a seed out of which a new plant may arise. But while the returning plant repeats the course of the previous year, the world creations move a step forward with each renewal.

From the results and achievements of these three stages of creation, the fourth, our present world, has been born. The three kingdoms of nature in their different degrees of evolution are expressions of these three world creations that preceded and prepared our earthly world.

The first stage of creation in the above book is called the world of "Old Saturn." It plants the first germs of the *physical nature* of man and attains to the degree of evolution of the mineral. Out of spiritual forms of existence it condenses only as far as warmth, and rises again into the spiritual.

After the first stage follows the world of the "Old Sun." In it the evolving image of man receives, from the spiritual world substance, the gift of the superphysical *formative forces of life*. These lift it to the plant-like stage of evolution (though here we must not think of present-day plants). The Old Sun is a creation of light, even as Old Saturn is a creation of warmth. Physically, it condenses to the air-like. At the end it ascends with its fruits again into the spiritual.

These fruits become the seed of the third world, that of the "Old Moon," which receives, as a new attribute, the quality of soul. This permeates the evolving image of man, raising it to the level of animal nature (which has nothing to do with the present animals, but is on a higher level). To the light-nature of the Old Sun is added tone, or sound, as a new quality. The world condenses to a fluid condition.

18

Out of this third stage, through a metamorphosis of all its results into new germs by passing through a purely spiritual form of existence, arises the fourth stage, our earth and our present cosmos. Within it man comes to his full unfolding, as the spiritual world endows him with the spiritual essence that is able to grasp its own being: the ego.

Physically, the new creation is a condensation to the solid state. The existence of these four stages of creation is, according to anthroposophical research, the reason why we have four kingdoms of nature, four main components of the human constitution, and four conditions of matter. This may be summarized as follows:

Old Saturn	mineral existence	physical body	warmth
Old Sun	plant-like existence	formative force body	light, air
Old Moon	animal-like existence	soul nature	tone, fluid
Earth	human existence	ego	solid matter

The various kingdoms of nature have their own distinctive principles. The characteristic of the mineral kingdom is the mineral body permeated by purely physical laws. The totality of the mineral is the world of death. The essential part of the plant, on the other hand, is the body of formative forces, the etheric body, the principle of flowing vigorous life. But the plant also has a physical body. This is built up from the substances found in lifeless nature but, to the extent that these are suffused by the etheric body, they escape from the dead mineral laws. The etheric forces give life to the plant body. The plant is thus a two-membered being, containing a physical and an etheric principle. The animal, too, has a physical body, which contains the same substances that we find in the mineral and plant kingdoms. It has, like the plant, a body of etheric, or formative forces, which conscripts the lifeless mineral substance into the service of life. The animal, however, presents not only the phenomena of life and death as does every living being; it also has waking and sleeping. It has sensation; it can experience pleasure and pain. A soul nature pervades it. This makes

19

the animal far more complete within itself. The plant is a whole being only in conjunction with the star-world, it has an "open structure." The animal is, by itself and because of the world of its inner organs, what the plant can become only together with the world of the stars (astra). It forms its body so that it can manage its sentient nature. To this soul-like quality the name "astral body" can properly be applied. The animal is entirely a creature of its instincts and of the organs formed by this third principle, the astral body. "It is instructed by its organs," said Goethe. Man, however, controls and instructs his astrality. He informs his organs (sees through illusions, for example). He harmonizes his instincts. For in him there is not only a soul nature but also spirit, self-conscious spirit that can grasp and recognize itself, ego. Thus man has four members: physical body, etheric body, astral body, and ego.

We find this presented with mathematical clarity by Rudolf Steiner in his first fundamental books, which appeared shortly after the turn of the century. One of these is *Theosophy, an Introduction to the Supersensible Knowledge of the World and the Destination of Man.** These books also show how one can advance to spiritual perception of the etheric body, astral body, and ego, by developing higher abilities that are latent in all of us. A sign that the time is ripe for us to grasp the fourfold essence of nature may be the so-called "layer" theory of Nikolai Hartmann, which speaks of inorganic, organic, psychic, and spiritual strata in order to grasp the whole of earth life and to distinguish between the various kingdoms of nature.

The earth has recapitulated, in an abbreviated form adapted to the new impulses of this fourth period, the "cosmic past" of the three preceding stages. Haeckel's biogenetic law, according to which the individual evolution recapitulates the evolution of the species, is valid also for the evolution of our solar system. In its gradual contraction out of the primeval planetary mist, our solar system repeated the conditions of the Old Saturn, Sun, and Moon creations. It contracted to the present Mars orbit. In the process of crossing this orbit, the creative impulses peculiar to the earth became increasingly decisive, until finally the middle period of earth evolution was reached. The first half of this evolution may therefore be designated the "Mars condition."

---

* Anthroposophic Press, Spring Valley, N.Y.

Alongside what is fully present, however, the evolving earth contains not only its own past but also the germs of the future. The future is being prepared by what is active within the earth. *Genesis* gives a view of the world's past; *Apocalypse* gives a view of its future. In new and future world evolutions these germs will achieve their fulfilment. In anthroposophy, these evolutionary phases, for reasons whose exposition would lead us too far afield, have received the names of Mercury, Jupiter, and Venus.*

Saturn, Sun, Moon, Mars, Mercury, Jupiter, Venus, as incarnations of the world process, as the great epochs of time in this process, have entered the course of each week as a "holy revealed mystery."

But let us return to the metals. Can we find, not in an indirect or metaphorical sense, but in their own attributes, indications of the secrets of time and space?

Many things will result from the individual descriptions of the seven metals that follow. Some general points may be listed here.

*   *   *

In a lecture cycle for physicians printed under the title *Spiritual Science and Medicine*, Rudolf Steiner points out that a study of the metals, if pursued as there suggested, would reveal such connections that one would have to correlate the metals and planets as follows (assuming in each case that the planetary influences have not been disturbed by other factors):

> lead to Saturn influences,
> tin to Jupiter influences,
> iron to Mars influences,
> copper to Venus influences,
> mercury to Mercury influences,
> silver to Moon influences.

But these undisturbed influences must be thought of as having existed in the earliest epochs of earth evolution when the entire earth was more

---

* The present planets bearing these names are connected with spiritual facts that justify this nomenclature. Those interested will find a thorough explanation in the last chapter of Rudolf Steiner's *An Outline of Occult Science*, and important elaborations in his numerous other works. They also describe the future development of man.

intensely permeated with the life and vigor of the cosmic formative forces and was still united with other heavenly bodies (or these had just separated themselves from it). Today such influences can no longer be expected.

Rudolf Steiner's indications go still further. He shows that ample opportunities existed for the formation of other substances, since various other planetary influences might have clashed or competed with the above-mentioned ones so that, for example, the lines of Mars effects might intertwine with those of the Saturn effects, etc. Thus, the less typical metals have arisen.

An unbiased observation of all metals must admit that there are among them unequivocal characteristic types, such as our seven metals. But others appear as something non-unified, a mixture of various types. In this way thallium reminds us in many ways of lead, in other ways of silver. Lithium is an alkaline metal, but has many features in common with alkaline earths. Palladium in some of its attributes reminds us of iron, in others of silver. In our heptagonal circle these relationships could be expressed by diagonals between two metal points. In all, twenty-one such "dual-component" metals are possible. Chapter 19 of *Spiritual Science and Medicine* points to a still more complicated interrelationship. It explains how the "antimonizing" force (described there in medical connections) would correspond to an interworking of Mercury, Venus, and Moon. Were these planets to work not individually, as copper or mercury or silver, but all together, their combined effect would be the same as that of antimony. In our heptagon such combined effects could be symbolically expressed by the triangles between the corresponding metal points. Altogether, thirty-five of such "tri-component" metals are possible.

Can we now show connections, spatially at first, between metals and planets? These would have to be the results of an exceptional combination of cosmic influences and aspects. Let us start with something reasonably familiar.

That biological processes run parallel with cosmic effects and planetary aspects has been noted with increasing frequency as research turns more and more to this field. Let us bear in mind, for example, the agreement between certain growth rhythms and sun-spot periods, as shown in the annual tree rings; how a certain human blood reaction coincides with sunrise or with the moment of an eclipse; how the activities of certain sea animals coincide with a particular moon position, as in the well-known Palolo worm; the favorable or unfavorable annual yields of certain forest trees in accordance with the movements of the related planets.

But such relationships reach even into the sphere of the lifeless. Colloidal chemical processes are shown to be dependent upon processes in our earth's surroundings. Of special significance here are the investigations of L. Kolisko, who justly titled the results of her researches, *Influence of the Stars on Earthly Substances.* This scientist succeeded in showing that a large series of metallic solutions, reacting upon one another and undergoing complicated colloidal-chemical changes, were sensitive to certain cosmic influences. Since 1924 she has published several studies with a wealth of photographic material.*

The photographs in Figure 5 are the result of a re-check (made by Theodor Schwenk in the research laboratories of the Weleda Company) of the Kolisko phenomena, in this case successive observations of the Saturn-Mars conjunction on 30 November 1949.

One percent solutions of iron sulphate, lead nitrate, and silver nitrate were poured in this order into flat round glass containers measuring about 11 cm. in diameter. Immediately after proper mixing, a strip of filter paper bent into cylindrical shape of about 7 cm. diameter was placed in each container, each cylinder with its joining slit toward the north. For several days before the conjunction, the experiments were made at the same hours and with the same arrangements, and also for several days after. In this arrangement the liquid rises in the paper cylinders, a strong reaction with precipitation of metallic silver and lead sulphate having already occurred during the mixing in the containers. Stimulated by constant changes in concentration, the liquid continues to rise and to react, and in so doing it forms characteristic figures. Around seeds of silver, additional silver separates along the lines of the upward stream. Between the tiny silver crystals, sulphate of lead precipitates. The totality of these rising and precipitating processes yields a characteristic picture.

* Leading titles are: *Silver and the Moon, Jupiter and Tin, Sun Eclipses and the Course of Gold and Silver Reactions, Saturn-Sun and Saturn-Moon Conjunctions and the Eclipse of Saturn by the Moon in Connection with Lead-Silver-Salt Reactions.*

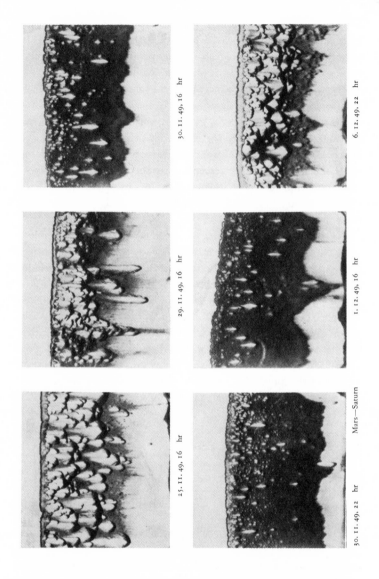

30. 11. 49, 16 hr

6. 12. 49, 22 hr

29. 11. 49, 16 hr

1. 12. 49, 16 hr

25. 11. 49, 16 hr

Mars—Saturn

30. 11. 49, 22 hr

24

Repeating these experiments at definite intervals of time, we find them presenting a constant type, until suddenly everything changes. The change in our case occurs when Mars passes close to Saturn. In the photographic series the pictures are still normal on November 25th, and also on the 29th. But on the day of the conjunction, November 30th, at 4:00 p.m., a completely unusual picture suddenly appears. The broad, heavy "lead formations" become pointed and narrow and are reduced in number; a strong blackening of the background occurs. Until the following day the reactions continue in this abnormal fashion, then slowly return to their normal condition. (For photographing purposes the paper cylinders have been unrolled.)

*   *   *

Can we now, in the properties of the metals, find signs of an ordering in time, having to do with phases of the earth evolution? To be sure, the arrangement of our heptagon results purely from certain traits of the metals themselves. Also, we have described certain phenomena implying that these metals possess a cosmic aspect in addition to the earthly, and especially that each of the metals is connected with a certain planet. Further, we have shown connections between evolutionary processes in the cosmos and the formation of certain planetary conditions. But have the metals something to do with the planets, not now as *spatially* oriented fields of force, but as phases in the evolution of our earth, that is, as formations oriented in *time?* Do the properties of the metals themselves justify the sequence shown in our obtuse heptagon (Figure 4)?

Such indications can be found, although they are of a delicate kind. In working through the next chapter of this book, on the nature of lead, the reader will note that lead behaves in a strange way toward warmth. Any increase or decrease of warmth affects it intensely. In the human body lead, as a therapeutic substance, is related to the organ-forming processes in which the ego intensifies or lessens the organic warmth. This points to the Saturnian aspect of its nature.

Gold, the sun metal, not only reacts in a definite manner toward the influence of the sun (see the investigations of Kolisko), but is to an unusual degree a metal of the light-world. No other metal can appear, as a metal, so radiantly colorful. It comes in shining golden yellow, in the bluish plant-green of the finest gold leaf, in the purple, red, violet, blue, and indigo of the colloidal gold solutions. The flashing up of the light nature is a characteristic of the Old Sun world.

Silver reveals in a particular way the world of sound and tone, and is also closely related to the nature of fluids. More about this will be found in the chapter on silver. But the ordering power of sound and the emergence of the fluid condition were the characteristics of the world of the Old Moon.

We have termed the first half of the earth evolution the Mars condition. The Mars metal, iron, is in a certain respect the most important metal in the material formation of the solid earth. The earth contains iron everywhere in considerable quantities, about 5% on the average. It is *the* metal of earth life. The first breath taken by everything newborn on earth is accomplished by the power of iron. The plants, too, turn green only if they find iron in the soil. Mineral, plant, animal, and man would be unthinkable in their present form without iron.

Mercury is not solid. It has retained something youthful, a fluid mobility that the other metals have already lost. It has maintained itself in this as yet incomplete condition despite its great density, its high compound weight. For this reason the second, as yet uncompleted, half of the earth development has been designated the Mercury condition.

Of tin and copper, Jupiter and Venus, we dare not speak in these connections. It is more difficult to find such phenomena the more distant the future toward which they point. What has already become, even what is in the immediate process of becoming, has condensed to sense perceptible reality, has imprinted itself on matter. What is to become in the future still lies in the supersensible. We shall search for it in vain with our senses.

* * *

In the preceding pages we have endeavored to give an introduction and to justify the order in which the single metals will be dealt with. Lead, tin, iron, gold, copper, mercury, silver, will be presented. To begin with we will describe each metal in its relationship to the earth, and its location, important ores, distribution over the earth. Then we will look at the pure metal itself. What it means for the plant world and for the animal nature will follow. Lastly, we will try to show its activity in man. Glimpses of its role in civilization will be included. Thus there will be no tracing of effects to causes, but from the diversity confronting our senses we will ascend to the essence, which at first emerges as an image. Sketches of the inner nature of the metals are ventured here. The author knows they are imperfect, but believes that an effort in this direction will link us more profoundly with the essence of the metallic. In the magnificent closing chapter of his *Theory of Color*, Goethe progressed from the sense perceptible to the moral effects of color, and thus succeeded in following the light in its activity from the physical to the biological, then to the soul realm, and finally to the spiritual. A perspective into these moral spheres will not be lacking in this work, such as opens up when we discover, for example, how lead is connected with impulses of consciousness, iron with those of courage, etc. For the mysteries of nature always find their highest forms in man.

III

LEAD

Dense and sluggish; soft and pliable, yet of poor malleability and therefore easily broken when rolled or stretched; melted by a mere candle flame; of a quickly fading lustre—this is lead, a chief representative of the baser heavy metals. The ways of the earth grip and affect this metal with ease; it can hardly defend itself. Hence, nowhere in nature do we find pure native metallic lead to any worthwhile extent. We must turn to lead ores if we are to begin in accordance with nature—above all, to lead sulphide, lead glance (galena). This ore will give us a bird's eye view of what, in its totality, we might call the "visible lead organ" of the earth.

Lead deposits are found principally in North America, Europe (Spain, Germany, Belgium), and Australia, decreasing in that order. Asia and Africa lag far behind. The figures* for production in 1938 will confirm this:

America (North and South)	52%
Europe	20%
Africa	4%
Asia	9%
Australia	15%

The South American production being comparatively small, however, the

* Berg-Friedensburg, *Lead and Zinc.* Stuttgart, 1950.

lion's share goes to North America. Thus lead favors the northern hemisphere, and forms a distinct "center of gravity" in the West, a lead equilibrium in Europe, and lead poverty in the East.

Strange to say, this natural distribution corresponds to the approximate needs of the respective civilizations: more lead is needed and utilized in the West, less in the East. The Western peoples live upon soils rich in lead and this civilization has augmented the possibilities for certain diseases, notably the aging and hardening processes, against which lead may be a good remedy, as we shall see at the conclusion of this chapter.

Natural Forms of Lead; Lead Ores.

Lead glance or galena, the most important lead ore, shows, in its specific gravity and its sparkling though sombre lustre, a dense and gross materiality that we immediately recognise as metallic. But who would suspect it of containing sulphur? This element—ever mobile, bright, open to all exterior influences, sensitive to everything that touches it—has been overwhelmed and paralyzed by lead. Of *its* nature nothing more remains visible in lead glance.

Ancient, rigid, hardly changeable, the deposits of galena lie in the parent rock. Toward the depths they mostly turn into zinc blende (zinc sulphide), which is lighter and appears much less metallic. Lead and zinc are thus strangely bound up with each other; lead glance preponderates in the upper levels, whereas the heaviness decreases and the lightness increases the deeper we descend. The lead impulse disappears into zinc in the depths. Because of this some great lead mines, such as Broken Hill in Australia, have in the course of time become important zinc mines.

But still another metal must be borne in mind whenever we speak of lead. All lead glance is delicately interspersed with silver glance (argentite). Also, lead ore veins are often intertwined with veins of pure silver ores, such as the fahl-ores or tetrahedrites. Nature bridles, as it were, the force of lead with a silver rein. Here the physician who would allow nature to teach him, in the sense of Paracelsus, can learn from it a process that may help man to defend himself against the harmful influences of lead.

The parent rock and the gangue, the way the ores lie and the company they keep, show that lead must be placed between the processes of silica and lime. The large lead deposits are frequently embedded in limestone, and the veins contain either calcite of carbonic minerals resembling it, such as manganese spar, siderite, or heavy spar (barite), or they contain quartz.

Lead sulphide (galena) puts the other lead ores considerably in the background. Nevertheless, they remain of interest to the physician and the pharmacist, for they show how lead is related to the most diverse processes in nature with which, in turn, certain processes in man may be associated.

The second most important ore is cerussite (white lead ore), also called lead spar or lead carbonate. We find it especially in the Congo. It is distinguished by beautiful crystal clusters growing in a confused lattice work of beams, bars, and leaves. These remind us of the forms in which the fine calcareous travecula arrange themselves in the heads of bones. Cerussite crystals are similar in shape to aragonite (calcium carbonate), which crystallises from hot solutions. An affinity with warmth emerges here, to which many another relationship will be added.

Next we mention green lead ore, pyromorphite, chlorophosphate of lead. This mineral is of the same chemical structure as apatite, revealing a further relationship between the lead processes and the lime processes, for apatite is a natural phosphate of lime. On the other hand apatite, with which some igneous rocks such as basalt are associated, frequently contains small quantities of lead.

Thus, for our spiritual eye, nature constantly draws lines from lead to limestone, to calcium carbonate as well as to calcium phosphate. We find these lines again in human bone formation.

The following are counted among the rare lead ores: massicot (lead monoxide) corresponds to artificially produced red lead; anglesite has brilliant colorless crystals of lead sulphate; red lead ore, chromate of lead, forms brilliant hyacinth-red crystals; phosgenite is a light-green chlorocarbonate of lead; resinite or lead gum is a yellow-reddish, grapelike, basic phosphate of lead containing aluminum oxide.

All of the above ores, with the exception of galena, are not primarily formations, but have derived from galena wherever the primary deposits have been exposed to the atmosphere or to the action of rocks containing phosphates. Lead has thereby been released from its connections with sulphur and carried over into the activity of carbonic acid, phosphoric acid, or oxygen. The primary processes of lead ore formation were obviously those in which, during primeval times, lead subdued the sulphur processes and thrust them down into the depths. Sulphur is by nature an atmospheric element. It is really a gas that has become solid through its frequent inner metamorphoses, but is straining to return to the atmosphere at every opportunity. It can be thought of only in conjunction with the air into which it dissolves so easily when it burns. But it is also an organic element, a component of all the proteins. In the mineral world it is fixed only by lime and by the metals. It is really at home in the organic. There it acts as an enlivening factor, accelerating the metabolism. Lead, as we shall see later, paralyzes the living regenerative metabolic processes, pushing them into the direction of hardening, sclerosis. In chronic lead poisoning, we try to rid the body of lead with energetic sulphur cures. This interplay between sulphur and lead, observable to this day in the bodies of animate beings, must have had its macrocosmic primal form in the processes that removed and mineralized the excess sulphur. We will meet this process often later. Other metals took part in this, but lead to an especially marked degree.

The ores here mentioned represent only the crudely material aspect of lead's nature. In tenuous forms it "homeopathically" pervades the whole of the earth. We find it in the soil; in the roots, leaves, and fruits of plants; in the bodies of animals; in the human body. We shall say more about this later.

LEAD AND WARMTH.

Much of the nature of lead is revealed in its varied relations to warmth. In the furnace the metallic lead is easily melted out of lead glance after removal of the sulphur. The heat necessary for its liberation is slight. How much more energy is required to extract iron from its ores! When lead, in the remote past, combined with sulphur to form lead sulphide, it obviously did not depart far from its metallic nature. It is akin to warmth in its metallic properties. It melts in a candle flame, at 327° C., and vaporizes at a temperature where many metals only begin to melt (1555° C.). Lead is extremely soft, so that it can be scratched with a fingernail. It is easily rolled to paper thinness or drawn into thin wire. No strong formative forces penetrate it; it remains plastic and ductile. Drawn across paper, it rubs off as a line.* It is easily deformed. A lead wire breaks under the slightest stress. It is, as it were, metallic clay or wax, quite capable of assuming or imitating strange forms, but unable to retain them.

Lead expands vigorously when heated and contracts strongly when cooled. It responds intensely to warmth impulses by expanding into space or withdrawing from it more powerfully than any other genuine metal. Furthermore, lead is an exceptionally bad heat conductor; a bar of lead can be held in the hand at one end, while the other end is already melting in a flame. The heat streams slowly from the hot end to the cold; it bogs down, stagnates, and becomes a marsh. Each particle of lead sucks in the heat, but only hesitatingly passes it on to the next. Think of a number of blocks of similar size made out of different metals; now fix or "model" into each block a "warmth structure" by differentiated heating, creating a "warmth organization." This will maintain itself longest in lead, while in silver it will most quickly blend and become uniform.

To electricity, too, lead offers great resistance, which disappears only after exceptionally intense cooling, for example by means of liquid helium, although then the resistance vanishes completely. With regard to electrical conductivity our metals may be grouped as follows:

Lead	Tin	Iron	Gold	Copper	Quicksilver**	Silver***
4.8	8.8	10.0	41.3	57.2		61.4

The same is true for the conductivity of heat. The sequence in this connection is already known to us. It corresponds to the acute-angled, seven-pointed star (Figure 3) in the preceding chapter. This arrangement now emerges in a new way, confirming the fact that it is indeed the "signature" of these metals. Quicksilver in this sequence must be considered as something apart, since it is liquid. It is a bad conductor of heat and electricity. But this immediately changes if we compare solid mercury with the six other metals. Solid mercury is a good conductor.

All the properties of lead described here are assumed by the other

---

* The lead pencils of former times were actually made of lead.
** Quicksilver, being a liquid metal, is here omitted.
*** The figures indicate the number of meters of wire, with a cross-section of 1 mm.², that produce a resistance of 1 Ohm.

metals when heated just short of their melting points. Thus lead, at its normal temperature, behaves as though it were quite close to melting. Although it is solid under normal conditions, it is so strongly connected with warmth that it behaves as though it were constantly at the extreme edge of solidity.

LEAD AND LIGHT.

The freshly cut metal shows a bluish luster, which soon grows dim when exposed to the air. This color is supposed to have given the metal its name.* Accordingly, lead would simply mean "bluish." *Mbliwom* an old German word-stem related to this, forms a bridge to the Greek *molybdos*.

Some ores, many lead salts, and the so-called lead glasses (lead silicate) have a different relationship to light. Because of their exceptional capacity to refract light, they produce a brilliant play of light rays. Incoming light meets with strong resistance. Moreover, they have a high ability to disperse light. Beams of light in darkness, when moving through such media containing lead, are subject to an inordinately high pressure of light and darkness on each other, resulting in the formation of unusually broad and strong colored edges. Lead glass, cut like jewels, therefore resembles the "pure water" of the diamond, thanks to its high capacity for light refraction and to its "strong fire" because of its great light dispersal. The better known lead pigments owe their intense radiance and their high capacity to these faculties. These pigments allow no background to shine through; they "cover" it and also protect it, as a skin protects a living being.

It is interesting to compare lead with silver in its behavior toward light. The two metals appear here as true polarities. Silver salts are directly light sensitive; the light releases the metal from its salts, but at the same time the metal may itself become colored, according to the color of the light it is exposed to, thus assuming every hue of the spectrum. Lead salts are not changed by light; they change *it* by uniting it intensely with darkness in their own substance, yielding thus the saturated and stable colors as a per-

---

* The German name for lead is *Blei,* which is said to be related to blue (*blau* in German).

manent embodiment of this union. In silver salts, light means transformation of matter; in lead salts, light is a sense phenomenon.

SOUND AND CHEMICAL PROPERTIES.

These two realms are here treated together because spiritual-scientific research has shown that the forces revealed in chemical processes are the same as those that appear in the phenomena of sound. In the liquid condition, which is its proper element, tone brings order and harmony into the chemical processes of the transformation of matter, while in the air it expresses itself as the forming, ordering element of the world of sound. It would exceed the scope of this book were we here to show how chemistry is, as it were, the inner music of the world of matter, the chemist an explorer of the counterpoint of that world, and how, conversely, certain inner qualities of matter become outwardly audible through sound. We refer the reader to the basic introductory work of Dr. G. Wachsmuth.*

Lead is not a sounding metal. Lead strings on a violin, a leaden flute or bell, would be unthinkable. The tone would be smothered in dullness.

Lead lacks the inner structure for the harmonious, vibrative response needed for tone. Only cooling with liquid air gives a leaden bell the ability to ring, or alloying it with calcium, the metal of lime! When we touch a sounding object with lead, the sound immediately congeals, all vibrations cease. (For this reason skyscrapers are often built upon a base of lead to check the vibrations of the foundation.) The chemical reaction of lead, too, is such that although it may be "touched" by all possible agencies, it causes the action of chemical forces to become stiff and paralyzed. It combines readily with the elements, with acids and alkalis, down to the weakest of them. In this, its "base" nature comes out. Mere exposure to oxygen or to the carbon dioxide of damp air destroys the shining metallic surface. Thus it cannot resist chemical influences. Yet as soon as it is taken hold of and brought into a combination, it forms something insoluble, something that precipitates heavily and precludes further chemical action. No other metal is so inclined to produce insoluble compounds. Hence there are few soluble lead salts. It literally "calcifies" in its chemis-

---

* *The Etheric Formative Forces in Cosmos, Earth and Man.*

try. This expression is all the more justified, since in its entire chemical behavior lead shows a strong resemblance to calcium and the metals of the calcium group, such as barium. Natural sulphate of lead (anglesite) is of similar structure (isomorphic) to heavy spar, the natural sulphate of barium; cerussite (carbonate of lead) has the same relation to aragonite (calcium carbonate). Like calcium, lead is used with silicic acid in making glass. Hence the possibility of melting out the previously mentioned lead glass or crystal.

This inclination toward lime shows lead to be a bi-valent element. (The metals of the calcium group are all bi-valent.) Like lime, lead also combines with sugar to form saccharates, a peculiarity that we can properly evaluate only when we consider the lead processes in man. But there are also fourfold "tetra-valent" lead combinations. These resemble silica, a substance which, in nature, is diametrically opposed to calcium. Thus lead tetrachloride resembles silicon tetrachloride. The "lead chemistry" draws its lines between lime and silica.

LEAD AND RADIOACTIVE PROCESSES.

The earth contracted out of the cosmos to its present state in a vast process of condensation. This led to the formation of substances of extreme density, representing a kind of culmination of the solidifying process. But earth evolution has, as we have said, already passed its middle point, and we are faced with the first traces of "devolution," of decomposition. These may be seen not only in the outer face of the earth, which according to the great geologist Suess* is actually a gigantic heap of wreckage, but also in processes stirring in the depths of matter, such as radioactive disintegration. Uranium and thorium, which are substances of the highest density, are also starting points for decomposing processes, for the disappearance of matter. The stages of radioactive decomposition emanating from them show constant transmutation of matter into other elements, where on the one hand matter disappears, is destroyed, and on the other hand non-material types of radiation, imponderables, are released. Basic substances of lesser density are thereby formed as transitions in the

---

* E. Suess, *The Face of the Earth.*

disintegration, among them repeated variations (isotopes) of radioactive lead, until finally the entire process of radioactive decay comes to an end with the formation of non-radioactive lead—at least in the present phase of earth activity. (In contemplating these phenomena one is tempted to say that lead is actually not a substance but a condition, a latent process.) The older a radioactive ore, the more lead it must contain. Lead is the sign of its age.

Therefore we find lead in entirely different places than those previously mentioned, though in quantities that are insignificant when compared with the lead glance deposits. These are the localities containing uranium and thorium ores.

Lead has more natural radioactive varieties (isotopes) than any other element. It is especially closely connected with the radioactive processes in nature. Rudolf Steiner once characterized lead by saying, "Just consider where the strongest forces of decomposition are found in the earth; where radium occurs we find the strongest forces of decomposition. In lead the cosmos prepares a substance for itself in which to concentrate its most powerful splitting forces. By bringing lead into the human body, you place the body directly amid the processes of world disintegration." And elsewhere: "In lead we actually have an effective means of evoking the forces of decomposition." When we come to the lead processes in man, we shall elaborate further on this aspect of its activity.

LEAD IN THE SOCIAL SPHERE.

Lead was well known in early antiquity. The Phoenicians traded in it and the Greeks, for the sake of its silver, dug for it in Cyprus, in Rhodes, and in the famous mines of Laurium. The Romans, however, were the first to use it extensively. They used it for plumb bobs, catapult missiles, pipes, barrel staves, hairpins, and tokens for admission into the arena. With its help they extracted silver in refining processes. Water flowed into their homes through lead pipes, so that these people of sober intelligence, during a long part of their history constantly absorbed minute traces of lead in their drinking water. Lead in minimal doses strengthens the forces of consciousness. We will discuss this later.

Yellow oxide of lead (litharge), white lead (lead carbonate), and red lead (minium) were well known. The Roman sources of lead were, as they are today, predominantly in the west of the then known world, in Spain. In Titus's time 50,000 slaves toiled in the Spanish mines.

In the Middle Ages, Germany (at that time the most important mining country), Bohemia, and Hungary, were the main sources of lead. Lead production rose to a degree never before anticipated when the modern era began, and we must remember that it resulted from the unfolding of new forces of consciousness. In this age of discoveries and inventions we have taken hold of lead on a really grand scale. We need its malleability for the manufacture of plate and sheeting, from which all manner of containers are produced. It is used for bearings. Its easy fusibility makes it useful for solder and type metal. Though chemically susceptible, the resulting compounds are insoluble, forming stable, tough, solidly-adhering coatings, so that this metal has become indispensable to the chemical industry in the manufacture of boilers, coolers, vats, tank-cars for corrosive acids, etc. The oceanic cables are protected by lead coatings, and metals are protected from rust by lead paint. These examples of the many uses of lead may suffice. To list them all would fill a book.

Lead affects many vocations. Factory workers making lead paint or electrical condensers, house painters, typesetters, are in constant touch with this metal. The air in many cities is impregnated with fine lead vapors because of the use of lead tetraethyl as an "anti-knock" factor in gasoline, diminishing its inclination to explode too rapidly in the cylinders. Such close and many-sided connections with this Saturnian metal have their effects: known, misknown, and unknown, to use Goethe's expression. Known are the damaging influences of regular minute doses of substances containing lead, with the terrible results of chronic lead poisoning. Misknown are the alerting and consciousness-sharpening effects of the extremely fine quantities to which typesetters are exposed. Rudolf Steiner once pointed to the active role of typographers in modern labor movements, and explained it by their constant association with the lead in type. Lead in such minute doses, working over long periods of time, delicately augments certain processes of decomposition, which in turn make possible a strengthening of consciousness. Unknown, at least to a great extent, is what may eventually result from the inhaling of minute quantities of lead vapor in the air. The increased prevalence of lung cancer in the large cities of the West may be a first warning sign. Think of the fact that many millions of pounds a year are thus vaporized into the atmosphere, representing the fourth greatest factor in the total consumption of lead.

Lead and the Plant World.*

In small quantities lead is present in all plowland and in all parts of the plants that grow on it, although each species absorbs different quantities from the same soil. If the lead concentration reaches the 6th or 5th decimal potency (1:1,000,000 or 1:100,000), we find life inhibiting eff?cts, even dwarfing. The plant's breathing is physiologically hindered and the lead accumulates in the roots. Animals feeding upon such lead-rich plants may be poisoned by them. Maize, the chief grain of the lead-rich West (America) is least subject to damage, wheat the most, and young plants more so than older ones.

Interesting results have been obtained from experiments carried out along lines suggested by Rudolf Steiner. He had been asked whether it might be possible to bring certain formative forces connected with lead to visibility by infusing them into the life organization (etheric body) of a plant. In its physical manifestation, the plant is an image of super-physical formative life forces. It makes visible to the senses something of a super-sensible nature. The force structures of earth and cosmos working in a given region may be understood from the forms and colors of the plants that thrive there. Flowers with a vertical growth-axis normally have a regular shape, such as that of a star or bell; whereas in those with a more horizontal axis, the weight tends to produce a form with upper and lower lips. Now, in order to make the lead forces visible, plants were grown in air containing fine traces of a volatile compound, lead tetraethyl. After a time, the leaves curled into semi-cylindrical shape and died at the edges and tips. Simultaneously, cells of the stalk tissue changed into vegetative cones which—a strange sight in a calendula—began to sprout laterally, but soon died off. Points of hyper-active vitality had appeared, just as predicted.

---

* Compare Scharrer, *Biochemistry of the Trace Elements*.

LEAD AND THE HUMAN ORGANIZATION.*

If we follow the distribution of lead through the human body with the refined analytical methods now at our command, we find something similar to what we found in the earth. In extraordinarily high dilution, about the ninth decimal potency (one part per billion), lead is found everywhere. In certain organs, however, it accumulates in larger quantities, as if it were attracted by these organs. In such accumulations we find it in the bones, but also in pathological calcifications and ossifications such as gall stones, kidney stones, and bronchial calculi. A certain type of gall stones might actually be called "metal stones" since, in addition to calcium, strontium, and magnesium, they contain copper, zinc, often some silver, and always lead. As lead is concentrated in such organs, conversely it is diluted in other organs to the point of being no longer traceable, above all in the blood. At first this distribution is a mystery, as is the distribution of lead in the earth. But in man, this riddle begins to speak. For in this distribution there appears a most impressive relationship to warmth. In our description of inorganic phenomena, lead's sensitivity to the workings of warmth became apparent. Now in man we find that the organs that are most permeated with warmth, such as the blood, maintain their lead in extreme dilution and permit no coarse materialization. In their differentiation, the warmth relationships of our body display a true organization, which may be called a "warmth organization" ** (R. Steiner). In this the human spirit lives as a creative ego; indeed, this ego molds the entire physical corporeality in a certain way with the help of the warmth organization. An increase in warmth tends to form organs that have a dissolving activity; a reduction in warmth creates organs that tend toward the lifeless, are subject to solidification, and can therefore serve as organs of support. We have such organs primarily in the bones. Where this organic cooling off takes place, lead precipitates as matter. Bones, stone formations, etc., become rich in lead.

---

* Flury, *Manual of Pharmacology*; Karl Zink, *Physical World Riddles*; Gerlach & Gerlach, *Research and Progress*.

** The realm of every organ has a specific degree of heat. In this respect the "warmth organization" is just as widely differentiated as is the physical body in space. Additional information may be found in F. Husemann, *The Picture of Man*.

To these physiological effects we can add psychological and spiritual ones, forming a true "lead picture" drawn by man himself. A healthy person taking small doses of lead regularly and for a prolonged period soon develops a melancholy mood, followed by dizziness and severe headaches radiating from the back of the head to the forehead. The acuteness of his senses diminishes. Gradually physical changes appear, signs of chronic poisoning. The optic nerve degenerates to the point of blindness. The muscles, notably the extensor muscles, become sluggish, then stiff and lame, and finally deteriorate, while the muscle cells atrophy and an excess of connective tissue is formed in their place. As a consequence, the sufferer is no longer able to make any relaxing movements, only contracting ones. He becomes spastic, cramped. His joints ache. His pulse is weak, tense, rigid. His limbs are cold to the touch. The respiratory organs react with severe chronic coryza, laryngeal spasms, hoarseness, dry cough. Abnormal dryness in mouth and pharynx seizes the digestive sphere; gastro-intestinal catarrh, retention of gases, and chronic constipation join in; the metabolism weakens. The body contracts with painful cramps; spasms of the bladder and urine retention appear.

Relationships of Lead to the Members of the Human Constitution.

To understand the relationship of a natural substance to man, we must ask ourselves the following question in all seriousness. Which member of the human totality is it that can take hold of this substance? We have already indicated that the totality of man comprises four essential realms or members: the physical body, the etheric or formative forces body, the soul-like nature or astral body, and the kernel of our spirit nature, the ego. These four principles are active in every organ and function, though their participation varies in intensity. In certain organs and functions the physical principle predominates, in others the etheric, in others the astral, while in still others the ego prevails. If we discover the relation of a substance to one or the other of these human principles, we can follow its effects throughout the entire organism, down into the single organs where the principle involved exerts a decisive activity.*

---

* Dr. R. Steiner and Dr. I. Wegman, *Fundamentals of Therapy, An Extension of the Art of Healing through Spiritual Knowledge.*

Every natural substance radiates forces and qualities by which it asserts itself. Entering into an organic totality, such as man, it must be divested of its own forces and qualities so that it may be seized by the forces raying into it from this totality. It may be said that the importance of a substance for an organism depends on what this substance is "open" or accessible to. What has become fixed in it must be surmounted. Its *potentialities* must be called forth.

Is there a harmony between lead and the nature of the physical body? We have seen that under the influence of lead the body degenerates, becomes brittle, and breaks down. In the etheric body's realm of activity, the fluid organization, there occurs a drying up and a gradual paralysis of all life processes. The astral body, in the air organization formed by it, likewise shows pathological effects such as colics, gas retention, spasms, and it experiences these pathological processes with increased pain. Thus there is only a slight kinship between lead and these three human members; its dynamics do not harmonize with their impulses.

THE EGO ORGANIZATION AND LEAD PROCESSES.

The ego is a spiritual member. Its nature and activity can therefore be explored only in a spiritual way. From the results of his objective spiritual research, Rudolf Steiner describes the activity of this ego organization as follows:

> The ego organization lives entirely in states of warmth. It works on . . . substance by enhancing or lowering the states of warmth of a nascent organ. If the ego organization lowers the heat, inorganic materials enter the substance and a hardening process sets in. The basis is thus provided for the creation of bones. Salt substance is absorbed. If on the other hand the ego organization enhances the warmth, organs are produced whose characteristic action is to dissolve the organic substance, leading it over into a liquid or airy condition. Let us assume that the ego organization finds in an organism so little warmth that no adequate enhancement can take place in the organs requiring it. Organs whose proper functioning lies in the direction of a dissolving process will fall into a hardening activity. They assume in a morbid way the tendency which in the bones is healthy . . . The blood vessels are the organs which, for the reasons above

mentioned, may pass into a formative activity similar to that of the bones. We then have the calcifying disease of the arteries known as sclerosis. In this disease the ego organization is, in a certain way, driven out of the organs.*

In another passage of the same book the ego organization is characterized from still another aspect:

> In the sphere of material substance we can trace the ego organization by the presence of sugar. Where sugar is, there is the ego organization; wherever sugar appears, the ego organization emerges to direct the sub-human (vegetative, animal-like) corporeality towards the human. Sugar is . . . present in the blood. The sugar-containing blood, circulating through the body, pervades it with the ego organization.

Thus the ego reaches above all into the warmth processes of the body, shaping them into a "warmth organization." This maintains at every point of the body the warmth proper to that point. An increase in this degree of living warmth enhances the dissolving metabolic processes; a decrease favours the solidifying form-processes. At every point a healthy balance is maintained between inflammation and induration, between dissolving and congealing. Blood and bone are two polarities in this organizing activity.

Now let us return to lead. It has the relationship to warmth that we have described. It expands more on heating and contracts more on cooling than any other solid heavy metal. It clearly shows its inner relationship to calcium. Its salt formations precipitate out of the fluid condition, leading to separation and congealment. It has the ability to combine with sugar. Its qualities match the dynamics of the ego. Thus it maintains itself, once taken up into the organism, in highest dilution within the warmth-bearing blood, but thickens wherever the ego has consigned organs to solidification and rigidity by lowering the warmth. Lead contains decomposing hardening tendencies, but these are normally under the control of the forming impulses of the ego organization. They become pathological when they unfold a life of their own that supplants these impulses.

* *Fundamentals of Therapy*, Chapter 12.

42

LEAD AS A REMEDY.

From what has been said above we may gather hints as to how lead can become a remedy. For this purpose it must not be left in its natural state, but so prepared pharmaceutically that it may adjust itself to the intentions of the ego. It must be an instrument for the enhancement of these intentions whenever they are too weak.

The course of our life is essentially a gradual penetration of our physical and etheric natures by our soul and spirit. A definite equilibrium among the principles of our being must establish itself at every period of our lives. This begins at the very moment of incarnation. It is then that the physical body must be so shaped that it can carry the soul-spiritual principle of the individual inhabiting it. This soul-spiritual principle (ego and astral body) is at first so absorbed in this task that it cannot reveal itself in its true nature, but hides in the bodily garments while shaping them. Only in its imprint does it appear in the physical-corporeal, as "the spirit that is building itself a body." If this soul-spiritual can fashion the corporeality into only an imperfect expression of itself, one probable result will be certain childhood diseases. If the child's organism remains overly warm, overly liquid, for example, it may refuse to perform adequately the processes of solidification and calcium infiltration that are needed so that later the soul-spiritual principle can express itself in the physical form. The ego expresses itself in the bony system, the skeleton, as form, in the blood as dynamics. The upbuilding forces may become too active and reject the forming forces. Here lead can be an important remedy, introducing into the body hindering (catabolic) forces that will facilitate the incorporation of the ego and astral body.

With this "incorporation" sufficiently accomplished, there begins, slowly at first and then in growing measure, the opposite activity; the soul-spiritual principle, having filled the body with its own laws and made it its instrument, slowly and gradually frees itself from it and takes the upper hand. Now the soul-spiritual can come into its own, can assert its true nature. The first clear step along this road is the changing of teeth. With this, the child becomes ready for school. The soul-spiritual grasps itself ever more consciously; on coming of age it attains to self-consciousness and learns more and more to treat its body as an instrument.

These are, of course, beginning processes of "discarnation." We grow older, and finally old. Along the way, man's ego must become increasingly his guide. If we grow old in the right way, it becomes ever surer of its spirituality. Meanwhile the instrument, the body, grows more fragile. Degenerative products of a mineral kind permeate it; calcification reaches a certain point. Together with this mineralization, a certain cooling sets in. The burgeoning forces of life diminish, the body enters upon its autumn and winter. But as the functions fade out that were inaugurated by the astral body, for example, those that brought about our physical maturity, the astral forces are released. They must be taken hold of and governed by the ego, but now in a body-free way. If the ego continues to develop when no longer sustained by the body, the process of aging can be a fruitful stage in human life. Only then can the ascent of the spiritual compensate for the descent of the physical. If in this period of life the ego is unable to achieve a proper further development, if no true aging is accomplished, there results an unhealthy preponderance (hypertrophy) of the astral forces that are being released.

In the course of the years, the soul frees itself more and more from the constraint of the body into which it entered at the time of incarnation. Rapture, pain, and sorrow in the aged man no longer express themselves in stormy heaving of the breath or wild beating of the pulse. The soul becomes thoughtful and restrained, but also powerful, if it lets the ego grasp it and reshape it. Its contents are transformed; the understanding spirit ripens these contents into fruit that can later be transplanted for use in eternity.

But at this threshold man can come to grief. The liberated soul-force that rises from its confinement within the body, but is not carried over into the spiritual discipline of the ego, may become capricious, stubborn, obstinate, and full of illusions. Its forces may grow rank, hypertrophic. Since the spirit does not receive it into its own sphere, it now becomes a mischief-maker for the body from which it is estranged. What had been an unconscious constructive activity within the body during childhood is now, lacking the restraints of the ego, a destructive element running wild. The abnormal decay (catabolism) enhances the mineralizing tendencies in the organism. The body, weakened by age, is less and less able to maintain life in these substances, which through their own nature tend toward death. The solidifying tendency, healthy and proper in the bones, extends over the entire organism and deposits calcium, for example, in the walls of the arteries.

Lead, appropriately prepared, works as a decomposing agent on behalf of the ego organization. It combats the hypertrophically destructive tendencies of the astral. We should therefore find in it a remedy against arterio-sclerosis. This was indicated by Rudolf Steiner, who used lead for the first time for this purpose. The pharmaceutical preparation suggested by him is an intense permeation of lead with honey and sugar, by means of certain warmth processes. Thus prepared, lead can become to a high degree an instrument of the ego.

Another novel remedy is the use of red lead oxide (minimum) against alcoholism. In cosmic warmth the grapevine ripens its fruit, which is related, through its high sugar content, to the ego sphere in the blood. In fermentation this sugar is converted to alcohol, which enters the blood quickly and directly and produces a kind of counter-ego effect. The ego becomes powerless, losing the functions of its consciousness. The astral forces fall prey to intoxication, which becomes fixed in the astral body as a craving. The higher members of our nature become too strongly attached to the corporeality, deriving pleasure from it during intoxication. Lead subdues the hypertrophic forces of the astral body that express themselves in craving. It strengthens the forces of the ego organization and produces an antipathy to alcohol.

SILVER AS ANTAGONIST OF LEAD

In small doses, lead furthers the proper decomposing effects of the ego organization. If it works too strongly, the body will exhaust itself and waste away. Since the use of lead in our modern civilization is constantly spreading, the danger of creeping lead poisoning rises in proportion. In silver we have a means of counteracting this danger. Here again we must thank Rudolf Steiner for pointing out that certain silver preparations can counteract the effects of lead, although the familiar safety measures of the lead industries are, of course, not to be neglected. We have seen lead and silver as antagonists in nature. Now we find them again in the same role on a high scale in man. We pass, as it were, "the examination of nature" (Paracelsus) if we utilize silver as an antidote against lead.

* * *

Here the circle of our considerations closes. We began with the consideration of lead in nature, showed its effects in man, and then turned our gaze back to nature. If in so doing we have found a "higher nature" arising out of nature, we will have presented not a merely natural description of lead, but a picture of its essence.

IV

TIN

The distribution of tin over the earth follows laws entirely different from those of lead. The number of deposits is much smaller. As a pure metal it is extremely rare, appearing generally only as an ore, and predominantly in one ore form only, that of tin stone (cassiterite). The only other tin ore would be tin pyrites (stannite), which is much rarer. Thus rare deposits and few ores characterize tin.

Most tin comes from the Far East: the southwestern provinces of China, Burma, the Malay peninsula, and the tin islands of Banka and Billiton. The next largest deposits must be sought in the antipodes of these places, in Bolivia. The third place belongs to Nigeria in Africa and the fourth to Queensland, New South Wales, and Tasmania in Australia. After these follow the old European sources of tin, the "Tin Islands" (the Cassiterites of the ancients), Cornwall, and the tin mines of the Bohemian and Saxon mountains (Erzgebirge) that were so important during the Middle Ages. Surveying the areas that contain substantial deposits of tin ore we must definitely term it a "tropical" metal.

If we bisect the earth through its center with the ecliptic (the plane of its orbit around the sun), we obtain great circles that are inclined $23\frac{1}{2}$ degrees to the equator. Upon one such great circle, or grouped around it, we find the various deposits of tin arranged in a certain symmetry. This great circle runs right through the Bolivian tin country: La Paz, Ororo, Potosi. In Africa it crosses Nigeria and the important tin deposits there. In Asia, it runs through Burma, to the north and south of which lie the great Asiatic alluvial deposits. At an approximately equal distance to the south lie the Australian and Tasmanian deposits, and to the north, on the other side of the globe, the smaller Spanish-Portuguese, as well as the famous English and Bohemian-Saxon mines. Were the earth to rotate perpendicularly to its orbit around the sun (as Jupiter alone among the planets does), were the earth's axis to shift $23\frac{1}{2}$ degrees, and the north pole thus fall close to Great Bear Lake in Canada, there would be an accumulation of tin in the land areas around the equator, a tin girdle, so to speak, and there would be one band of tin each in the northern and the southern temperate zones.

The main tin areas of the world contribute to the total production as follows:

Asia (China, Burma, Malaya, Banka, Billiton) about one-half
South America (Bolivia)   "   one-fifth
Africa (Nigeria)   "   one-sixth
Australia   "   one-tenth
Europe   the remainder

Tin Stone (Cassiterite).

Tin stone, the most important tin mineral, is tin dioxide (SnO_2). It contains 78.6% tin. Thus the main ore is not a sulphur compound, as in lead, but an oxygen compound. In trying to understand tin we shall have to follow the ways of oxygen and not those of sulphur.

Tin stone makes no metallic impression such as lead glance (galena), but rather a jewel-like one, at least when it occurs in clear brilliant crystals. Most of the time, however, it shows brown or black discolorations, because of traces of iron. Despite its somewhat different crystal form (tetragonal instead of hexagonal), it has something rather quartz-like; great hardness (6–7), and a vivid, glassy lustre. Occasionally it appears in fibrous, ribboned form: wood tin. Whereas in our studies of lead we were soon confronted with lime, here we find a variety of connections with silicic acid. SnO_2 is tin stone, SiO_2 is silicic acid. We find the latter as rock crystal, but also in coarse, ribboned form, as agate, etc. We will notice further analogies between tin and silica when we consider the chemical properties of

48

our metal. But the friendship of tin stone toward silica already shows in its deposits. Tin occurs primarily in the oldest rock, the primeval granite, and this contact so transforms the granite as to make it into a species of its own, the so-called tin granite. Granite is the harmonious mixture of quartz, feldspar, and mica, but wherever it incorporates a vein of tin, the feldspar is destroyed and tin stone assumes its place. A special kind of mica, lithium mica, takes the place of the ordinary one. Simultaneously, the jewel-like nature appears: tourmaline, the fluorspars, topaz, and similar fluorine minerals accompany the tin veins. The adjacent rock tends to become "topazed." Topaz rock is formed. Apatite, the crystalline phosphate of lime, joins in. The granite is destroyed. Such processes point strongly to a relationship between tin and everything silicious.

TIN PYRITES (STANNITE).

This is the second important tin material, though far behind tin stone. It is a copper-iron-sulpho-stannate. Tin appears here as acid forming, but oxygen has been replaced by sulphur, although in this combination the sulphur has assumed an oxygen-like function. Tin pyrites is much softer than tin stone. It contains 29.6% copper, 13% iron, and 27.6% tin. Thus we have an ore in which iron and copper meet in the sphere of sulphur, at the same time permeating their activity with tin. This ore might consequently be useful to the physician desiring to direct the process of protein formation (always stimulated by sulphur) into areas affected by tin through the blood formation brought about by copper and iron. This might be of special significance in the sphere of the liver. Outwardly, tin pyrites looks like a metal, but is yellowish-brown like other pyrites. It is the chief ore in the Bolivian fields.

TIN AS A METAL.

Tin shines like silver, with a slightly yellowish hue. It is stable and looks almost like a precious metal. Attempting to describe it, we soon become involved in notable contradictions.

Tin is very soft, somewhat harder than lead, but softer than gold. If a rod of pure tin is bent, it emits a strange crackling noise, the so-called "tin cry." This is because its inner crystalline structure causes the bending crystals to shift against each other. If we apply a corrosive, this hidden crystal structure becomes visible. *Thus tin is formable and formed at the same time.* It melts readily, at no more than 232° C. But it must be heated to 2300° C. if it is to vaporize. Thus it has the tendency to reach the liquid condition quickly, but to leave it reluctantly. Let us compare it with the six other metals here under consideration:

	Melting point:	*Boiling point:*	*Difference:*
Lead	327	1555	1228
Tin	232	2300	2068
Iron	1530	3235	1705
Gold	1064	2677	1613
Copper	1083	2305	1222
Quicksilver	−39	357	396
Silver	961	2100	1139

Tin greatly surpasses all other metals in the spread between the melting and the boiling points.

If we heat soft ductile tin, it suddenly becomes hard and brittle, so that it can be pulverized. Heat strives to dissolve the solidity and inner form of all matter, to make it loose and soft, but tin resists these tendencies of heat. Conversely, when tin is cooled and we expect it to become gradually harder, denser, and more rigid, we are surprised to find that it takes on a light, loose, powdery structure. This transition begins at about 18° C., slowly at first, then faster and faster as the temperature sinks to −50° C. (Significantly, this is about the temperature at which quicksilver becomes solid.)

Thus tin occurs in three forms according to the degree of heat to which it is exposed. (a) Between 18 and 160° C. it appears as the well-known ductile metal with its silvery gloss, but already containing certain form tendencies that distinguish it from the other six metals. Gold, silver, iron, mercury, copper, lead, crystallize in the "regular" system, based on three perpendicular, equivalent axes. They generally crystallize in small octahedrons. Tin, however, crystallizes tetragonally with three perpendicular axes, one of which has a length of its own, while the other two are equal.

(b) When we heat tin above 160° C., it forms a new modification, rhomboid tin. The rhomboid system of crystals is based upon three perpendicular axes, each of which has its own length, which need not be equal to either of the others. Much greater differentiation relative to space is expressed herein. This rhomboid tin is brittle and harder. (c) On cooling below 18° C., the third modification sets in. Now the tin no longer has a metallic character, but is a gray amorphous powder much lighter in weight. (This powder has a structure of cubic regularity.) During cold winters the shining tin is inclined to turn into this gray dust, especially when it is scratched and some of this gray matter is rubbed into the scratches. Our ancestors discovered this to their dismay when the tin roofs of churches were subjected to prolonged periods of severe cold. The "sickening" of such tin begins with the formation of blisters that resemble plague sores and spread until all of the metal has disintegrated. This has been called the "tin plague," and is also known as the "museum sickness" because rare old pewter vessels or precious tin castings have succumbed to it. Through melting, "sick" tin can be converted back into "healthy" tin.

In this double opposition to the usual action of heat, tin shows a highly characteristic arbitrariness. It insists on maintaining its own form in the face of heat. It defends itself against heating as well as against cooling. It wants to remain in equilibrium between melting and congealing. It holds fast to the properties that produce an intermediate region of warmth, refusing to be softened by heat or frozen into shape by cold.

In these properties tin can be contrasted only with quicksilver. The two metals have the lowest melting points among our seven. Both have a distinct relationship to the liquid state. But tin contains as its hidden form, the crystal, quicksilver the drop.

Tin is a poor conductor of electricity and heat. Here it resembles lead and iron. On exceptionally deep cooling, at the temperature of liquid helium, it suddenly becomes "superconductive." All the metals pertaining to the planets "above the sun" are, as previously mentioned, bad conductors of electricity. Those belonging to the planets "below the sun" are good conductors.

Tin is the metal with most "isotopes." Next to it, in this respect, is its antagonist quicksilver.

51

Metal:	Iron	Copper	Silver	Tin	Quicksilver	Lead
Number of isotopes:*	2	2	2	11	7	4

CHEMICAL BEHAVIOR OF TIN.

Tin is a stable metal that may almost be called precious. It is therefore useful in the protection and coating of common metals (tin foil and plating). The preciousness of a metal may be established, among other means, by its electrical tension against a "standard solution" of its salt. From the figures thus obtained we have the electromotive series, whose zero point is established by the voltage of a hydrogen electrode. The voltages of our seven metals are:

Iron	Lead	Tin	Copper	Silver	Quick-silver	Gold
-0.43 V	-0.12 V	-0.10 V	$+0.33$ V	$+0.79$ V	$+0.86$ V	$+1.5$ V

Metals with a low voltage have a strong tendency to dissolve into salts, thereby easily losing their form and revealing their non-precious character. A metal in the electromotive series precipitates the metals to its right (in the series), and as a result itself enters the solution.

Again we have a separation of iron, lead, and tin from the other three metals: copper, silver and quicksilver. The two triangles of our seven-pointed star come to mind.

Tin forms two series of compounds. In the unstable bi-valent compounds it resembles in no wise the bi-valent lead compounds. It forms soluble compounds with most of the acids, thus showing none of the congealing tendencies of the chemistry of lead, with its desire to "calcify." But the bi-valent compounds strive energetically towards the tetra-valent condition. Here tin shows its kinship to the chemistry of silicic acid. Its chlo-

---

* Isotopes are variants of the same metal, but of different "atomic weights." These variants differ only minimally as to density, melting point, and compound weights. The "atomic weights" are whole numbers, and the mixtures of the various isotopes yield the fractional average atomic weights generally known.

rine and fluorine compounds resemble those of the silica compounds as no other metal compound does, especially among our seven metals. Tin dioxide combines with alkalis to form stannates similar to the silicates; there is a normal stannic acid and a metastannic acid, just as there are the corresponding silicic acids. The fluoro-stannates resemble the fluoro-silicates. This relation to silicic processes will have to be pursued further in the realm of the animate.

Tin mixes easily with other metals. Consequently, there are a great number of tin alloys. Its great value as a solder of many metals is due to this property. Many of the alloys make use of the pliability and softness of tin. A great variety of bearings and babbitt metals are so composed. Its easy melting is appreciated by the tin foundries; printers add tin to their type metal. Conversely, it is the forces of form and hardness hidden in tin that are called upon when the tough and solid bronze is produced from soft, pure copper. It invests the same metal with form and solidity in the casting of bells. Gongs, horns, tubes, organ pipes, contain tin for the same reason. The euphonious sound cannot emerge from copper only, because it lacks the strength of form. Tin calls it forth by the harmonious rhythmical tremor with which the union of copper and tin answers when the chord is struck. For in tin strength of form and plasticity of substance are ideally united.

TIN IN HISTORY.

Men acquired the use of tin in early ages. Tin jewelry has been found in old Persian tombs; the Babylonians and Egyptians knew it as a rare and valuable metal. In China it was known two thousand years before our era. In other areas there was an actual Bronze Age, which succeeded the Copper Age and carried over into the Iron Age. The Romans had their special artisans in tin. They were acquainted with tin foil and knew how to make mirrors from it. In Greece, too, in the fifth century B.C. tin was an article of trade. It came from far away, from the ends of the ancient world, from the "Cassiterites." From the Celtic world of the British Isles a stream of tin flowed into the Mediterranean. It was used for making weapons, axes, utensils of all kinds. Its name probably came from the Celtic *ystaen, sten, staen,* from which are derived the Germanic *Zinn,* the English *stanneries,* and the Roman *stannum*. In the Middle Ages the Bohemian and Saxon deposits were added to those of Cornwall. In the form of pewter, tin served for cutlery and common utensils. One ate from pewter plates, drank from pewter cups, used pewter candle holders. But also objects of art were cast from it. It served as a kind of poor man's silver for jewelry, boxes, jars, etc.

In modern times tin again serves the purpose of war in cannon and as a metal of technology. Tin salts are used in dye works and as a filler for silk. The European deposits having been exhausted or impoverished, tin for these purposes now comes mainly from Eastern Asia. It brings much gain to the seafaring nations. England, main source of tin during antiquity, now is the center of the tin trade. But the great conflicts and military involvements in the East during the last decades brought first the Bolivian and then the African sources into the foreground.

Tin in the Realm of the Living.

It is reported that certain plants like to grow on soils containing tin, for example, *Trientalis europ.,* a plant of the Primrose family, on the old ore dumps in the Bohemian mountains. As our methods of research grew more refined, it became possible to show traces of tin in the bodies of animals as well as man. French investigators discovered a remarkable distribution in cattle, horses, and sheep. The figures* for lungs, kidneys, blood, heart, brain, spleen, liver, skin, tongue muscle, and tongue mucosa are as follows:
We notice that the tin content of the liver is somewhat above that of the other inner organs. Toward the periphery, toward the spheres of activity of silicic acid in the skin, we find much more tin. The accumulation in the mucous membranes of the tongue, and even in the tongue muscle is extremely curious. We shall return to this at the conclusion of this chapter.

Tin and Man.

To understand the role played by tin in the human organization, be it in forming, in functioning, or in healing, what we find in nature will have to

---

* Scharrer, *Biochemistry of the Trace Elements.*

Lungs	0.98– 2.04	mg. per kilogram live weight
Kidneys	1.12– 1.78	"	"	"
Blood	1.25– 1.64	"	"	"
Heart	1.47– 2.42	"	"	"
Brain	2.4 – 3.0	"	"	"
Spleen	2.4 – 3.1	"	"	"
Liver	2.14– 3.73	"	"	"
Skin	6.20– 9.48	"	"	"
Tongue muscle	12.2 –16.5	"	"	"
Tongue mucosa	18.7 –26.11	"	"	"

be enlarged and transformed by spiritual-scientific observation of its effects in man. Here again, as in the case of lead, we must bear in mind that within the realm of the living the significance of a substance depends on what it allows itself to be *seized* by. What the substance is of itself is not significant, but what may happen to it through the totality that governs it. As Rudolf Steiner says, "The organism is not a combination of matter, but of activities." We will find the thread into the labyrinth if we observe substances with whose help the body carries out important processes and study how tin relates itself to such substances. Sulphur and silicic acid are such substances. Their paths through the inanimate as well as the animate are easily traced.

Sulphur is intimately related to protein in its composition and in its subsequent formative processes. Substances that are inclined toward the ways of sulphur in nature, such as antimony, silver, quicksilver, and copper, must likewise be followed in connection with the proteinaceous processes in the organism.

Silicic acid, however, has to do with shaping processes. It actually forms "the physical basis of the ego-organization." * We find silicic acid at the outer limits of the organization, in skin and hair, shutting off the workings and forces of outer nature from the inner organism. It helps create and maintain the boundary between outer and inner world. (Tin, as we have seen, follows in its footsteps.) Yet we find silicic acid also inside the organism, in the sheaths that enclose the organs. It sets up a second,

---

* See R. Steiner and I. Wegman, *Fundamentals of Therapy.*

inner, boundary to certain organic formative processes. Both spheres, the world without and the world of our inner organs, lie outside our consciousness. But between them is the realm in which we can unfold in the largest way those organs that develop our consciousness, that is, the organs and activities of the senses. The inner and outer sense activities need silicic acid. That we can perceive the world through our sense organs, and that the organs inside our body can "perceive" and experience each other (though in a manner of which we are scarcely conscious), for these purposes we need silicic acid. This is because silicic acid, by its own shaping tendencies, can become a fit instrument for the intentions of the ego organization. Silicic acid has only scanty relationships to sulphur, but the situation is quite different with oxygen. The same may be said of tin. We have already seen how tin follows the silicic acid processes in nature. In its plastic-colloidal form silicic acid, when combined with water, can swell up to an albumen-like, gelatinous mass. On the other hand, it presses irresistibly away from this condition and toward a crystalline form, such as finds its most beautiful development in the rock crystal. Here silicic acid, like a sense organ, is fully open and pervious to the cosmic forces, to light, to warmth.

Thus in tin, as in silicic acid, we find the polarity between the forces of plasticity and rigidity. Everywhere in the organism tin can aid the processes that establish the correct relation between the liquid and the solid. Such processes appear in the organism in manifold metamorphoses. During childhood, for example, the proper relation must be established in the head between the solid and the liquid for the brain and the cerebrospinal fluid in which the brain must swim. There must be neither hydrocephalus nor its opposite. Rudolf Steiner, in *Spiritual Science and Medicine*, actually spoke of the fact "that in this entire complex of infancy it is the design . . . to bring about the correct relation of solidity between the head and the soft parts," and that, "the same forces are active in this as in tin."

In the joints we discover a metamorphosis of the forces that produce a proper relation between the fluid and the solid in the head. Thus the joints are a sphere allied to the tin processes. The joints, too, are the area of equilibrium between two entirely different formative processes. It is the articular cartilage, covering the head of the bone, within which the true bone tissue resides. Between the heads of bones is fluid as the content of the capsular ligament. We know that in the embryo the entire bone is a cartilaginous formation. As growth continues the cartilage is overcome and replaced by bone formation that, proceeding from the center toward the extremities, leaves only the cartilage of the joints as a last remnant. Bone and cartilage are polar opposites because of their formative impulses. The structured, almost mineral bone matter stands in contrast to the watery, homogeneous, non-structured substance of the cartilage. In certain diseases of the joints this healthy relation is disturbed. The cartilage degenerates and the bone proliferates beyond its boundaries, occasionally splintering into the cartilage of the joint. We are confronted with the joint degeneration of arthritis.

An opposite picture is presented by the abnormal secretion of water in the capsular ligament, the effusions into a joint. Here the watery element predominates, escaping from the organizing life forces as it asserts itself beyond normal measure.

In both diseases tin is an effective remedy. Possessing softness and plasticity "outwardly," moderate hardness and forming forces "within," it represents the counter-process in nature that can summon up the healing processes in such cases.

The liver is, in a strange manner, an organic counterpart of the tin processes. The liver is primarily a plastic organ. It is really the largest gland in the body's fluid organization. From the digestive process it receives the liquefied food, while on the other hand special processes occur in the liver of densification and substance formation, such as the production of solid liver starch from the dissolved carbohydrates. To this rhythmical play between dissolving and condensing is added the special relationship of the liver to the warmth processes of the body. It is the warmest organ. This points to the fact that it absorbs, in a particular way, the working of the ego. As an organ of the fluid organization, its inner structure is strongly shaped by the etheric body. But the liver can also follow the impulses of the etheric body too vigorously. It can, as it were, overflow with vitality. In this case we must "bring to the organism something from the surrounding world that has an effect opposite to the action of the organ, something that can combat the latter. We must try to discover the external quickening impulses that correspond to those of the individual organs. . . . This we do when we combat the pathological liver activity . . . with the outer activity expressed in the metallic nature of tin."

The liver can also fall prey to the opposing activities. The etheric within it can be too weak, so that the physical becomes too strong. Disease aspects will then appear that tend toward condensation, hardening, and even cirrhotic changes. In both aspects tin has proved to be an excellent liver remedy. Its play between fluidity and form enables tin to intervene in the equilibrium between the etheric and the physical processes. Because of its peculiar counter-force against external warmth, it works on the ego organization so that this can also play into the equilibrium between the etheric and the physical.

* * *

Tin may be broadly described as a remedy that regulates opposing fields of force, on the one hand, those that appear in construction, growth, and swelling, and on the other hand those of solidification and drying up.

Where the liquid-plastic pole preponderates, with its construction, growth, and swelling (which may appear in the most varied ways according to constitution or age), tin in moderate or higher potencies will help. To bring form into a disordered structure, and to reabsorb and solidify what is too liquid, a tin therapy is useful in all spheres: in the head with hydrocephalus; in the middle region with exudative tuberculosis, pleurisy, and bronchitis; in the metabolic area with liver congestion, slowing of the bile flow, disturbances of resorption and elimination in the colon, colitis, hemorrhoids, and intestinal parasites. The same is true of eczemas due to liver malfunction and of all effusions of the serous membranes, such as meningitis, pericarditis, dropsy, and effusions in the joints.

Where the pole of excessive degeneration, of abnormal drying formations predominates, low potencies of tin are applicable. This will be possible in cases of microcephalus, cirrhotic forms of tuberculosis, cirrhosis of the liver, arthrosis deformans, gout, and also in Basedow's disease (exophthalmic goiter). Preparations of copper may supplement the tin in these cases.

At the beginning of this chapter we described briefly how tin has entered into granite, the silicated primal rock, in contrast to other metals that are at home mainly in the deeper-lying basic rocks or in the sedimentary and slate formations of the Paleozoic Age. Fluorine minerals have formed themselves in closest proximity. We find calcium fluoride, apatite, topaz (as precious stones), and frequently a development of topaz rock. The adjacent rock has become "topazed." Within the tin formations, there is thus a meeting of tin, fluorine, silica, lime (in calcium fluoride), topaz, and apatite. A strange community.

Now, as a counterpart to this mineral region and formation, we have an inner meeting of tin, fluorine, silica, and lime. This meeting takes place in the oral cavity, the mouth! The teeth, we know, are richer in fluorine than any other organs in the body. In them the fluorine process comes to rest, solidifying to the hardest formation in the body. Fluorine as matter is an enormously active element, the most active that we know. The body makes use of its dynamics in the transition of the metabolic life into the nimble activities of the limbs. It comes to rest in the form of tooth enamel. The sculptured form of the tooth evolves out of the ectoderm, a realm in which, as already mentioned, the silicic acid process is especially active. This sculpture is hardened and solidified by the fluorine, which combines with lime to an apatite-like formation. (Fluorine can volatilize silica, but is tamed by lime. This indicates its chemical behavior. Whoever writes a fluorine chemistry must keep it between silica and lime.) In tooth enamel we actually have apatite in the human sphere.

Facing the rock-hard teeth is an organ that lives wholly in fluids, the softest, most mobile organ in the mouth, the organ that constantly sculptures speech, the "speech limb," the tongue. In speaking we move constantly back and forth between form and fluidity. We now recall that modern investigation of the tin content of the various organs has shown that the one richest in tin is the tongue. In the mouth there is constant contact between the organ richest in fluorine and the one richest in tin. The tongue and its mucous membranes have been found to be particularly rich in tin. The tongue as a solid organ swims in the liquid of the oral cavity. This liquid acts in a way completely opposed to solidity; it has a dissolving action, thereby initiating the first steps of digestion. Here the carbohydrates in particular undergo a transformation. Starch is turned to sugar, a dissolving process that is reversed in the liver. In this way the plant starch becomes human starch. Tongue and liver may be considered polarities along the road of nutrition. Topaz, the third mineral mentioned in this connection, may likewise be used as a remedy. It is, according to Rudolf Steiner, related to the sense of taste.

So we find the geological triad of tin-fluorine-topaz in the human body, but strangely changed. Such flashes of insight granted by living nature challenge us to search for the primal connections between the extra-human and man. These connections have been fully described in modern times by Rudolf Steiner in such books as *An Outline of Occult Science*.

V

IRON

We cannot speak of iron in the same way as of lead or tin. Iron is so important and indispensable to all kingdoms of nature, as well as to man, that in it we become immediately and directly aware of what in other metals we must seek out as a hidden side of life.

Iron is an important constituent of the mineral kingdom. It permeates the earth everywhere, all rocks and soils. Among the seven metals it is the most abundant. According to Clarke, the earth's crust accessible to us contains the following percentages of our seven chief metals:

Iron	4.7%
Copper	0.01%
Lead	0.002%
Tin	0.0006%
Silver	0.000004%
Quicksilver	0.000003%
Gold	0.0000001%

To make these figures more pictorial, one long ton (1 million grams) of soil contains, on the average:

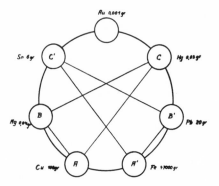

In the above figure the quantity decreases symmetrically along the lines of the two equal triangles ABC and A'B'C'. This regularity is brought out by our heptagon.

Comparing iron not only with the metallic but with all elements of the known earth, we find it to be the fourth most abundant. Oxygen, silica, and aluminum are the most abundant, in the order listed. Immediately after these follows iron. It is the only heavy metal so prevalent that we can ascribe to it, on the basis of material quantity alone, a role in the formation of the earth. There are whole mountains consisting mainly of iron ore, such as the one of magnetite ore near Kirunavara, Sweden, or the Iron Mountain in Styria.

Iron not only permeates rocks and soils everywhere, it also forms a great many important minerals. It is really the viceroy of the mineral kingdom. It forms these minerals and ores with the substances—oxygen, carbon dioxide, water, sulphur, phosphoric acid, silicic acid—that keep the most important earth processes at work and at the same time play the greatest imaginable role in the life of plant, animal, and man. Iron occupies the center of the stage. It cannot be overlooked or undervalued.

When we move from the mineral to the plant kingdom, we find this confirmed again. No chlorophyll can be formed without the cooperation of iron, nor can the plant be built up from the cosmos by the power of light. Therefore, every plant contains iron in its ash. In some plants such as anise, nettle, spinach, and the fruit of the water chestnut, iron is especially abundant. The evolution of the animal kingdom is equally unthinkable without iron. There could be no red blood, and consequently no breathing of air. Lower water creatures may breathe with "copper blood," but if life is to ascend to lungs and limbs, to a hard inner skeleton, to a sounding voice, and to a waking soul life, it must ascend also from copper breathing to iron breathing. In the plant's chlorophyll, iron is merely a catalyst and does not enter as a substance. It is needed in the environment of chlorophyll formation, but is not incorporated in it. In hemoglobin, however, it not only cooperates but enters right into the substance. The plant leaves iron "outside" as it were. The animal takes it "inside," into its blood.

But it is in human life that the iron process is really fundamental. It is built right into the central ego organ, the blood,* and thereby becomes an instrument not only of the waking soul life** but of the willing ego. Iron carries the mineral nature all the way up to the human blood, but the mineral nature is conquered, pervaded, and transformed by the fire of the blood. We will show later how this process is also the prototype of healing, how iron is the greatest remedy in the human realm.

In its fields of activity there is a convergence of the aeriform and the liquid, a meeting of oxygen, carbon dioxide, and water with breath and blood. Here we rediscover the patterns that iron, with its tendency to combine, has already drawn in inorganic nature, like hieroglyphs. In man these hieroglyphs become decipherable.

Without iron the rocks and soils would not be brown, green, yellow, or red; the plants would not be green; man would not be flesh-coloured. Thus the world, in all its kingdoms, owes much of its complexion to this metal.

THE IRON BELT OF THE EARTH.

As with lead and tin, let us try to trace the laws by which iron is distributed over the earth. Although iron is present everywhere, only in certain localities is it so abundant as to justify mining. When we survey the largest ore deposits, the North American, English, French, German, Russian,

---

* Rudolf Steiner: *The Occult Significance of Blood.*
** Consider, for example, the dreamy nature of "blood-poor" anemic young people suffering from a deficiency of red corpuscles.

northern Chinese, we see that they form a mighty belt around the earth. By comparison, the Brazilian, African, Indian, and Australian deposits are of lesser significance. Thus the ores are spread throughout the north temperate zone. They are oriented in the direction of the true axis of the earth and not, as in the case of tin, toward the axis of the ecliptic. Even more remarkable, carbon shares with iron in this arrangement. It is widely connected with iron in the processes of life. The great coal deposits follow the iron belt through all the northern continents. The temperate zone is the rhythmic middle region of the earth, pulsing with the liveliest weather changes, the rhythmical sequences of high and low. Here the seasons work in reciprocal harmony, the forces of the poles reaching down in winter as the tropical forces reach up in summer. If we look at the workings of the formative forces in our earth as described by G. Wachsmuth in *Earth and Man*, we find a true rhythmical system active in this zone. It is the chest or thorax of the earth, just as we may call the poles, the head, and the equator, the metabolic region. Here prevails the degree of warmth most salubrious for man; here the largest land-masses span the earth. Coal, iron, and the rhythmical region of the earth belong together; the evidence shows this clearly. But it is curious that iron ore and coal lie close together in just the regions where they are needed by the man who has attained a certain stage of consciousness, the man of the technological age, of the "fifth post-Atlantean culture." * Consider how far the tin deposits of Asia and Bolivia are, for example, from the coal needed for their smelting, and also from the nations that use tin.

PURE IRON.

Earthly Form. To the north of the iron ore belt mentioned above, the small island of Disko near Western Greenland is the only spot on earth that is known to contain pure iron. It occurs in great blocks, weighing many tons. In small quantities, and in almost microscopically fine distribution, pure iron may be found in some German basalt. The Disko iron, too, is found in the basalt of greenland.

Cosmic Form. Other than this, the only pure iron found on earth is me-

---

* Rudolf Steiner, *An Outline of Occult Science.*

teoric iron. This cosmic iron descends to earth uninterruptedly (in special abundance during autumn) in the meteor showers. We refer not only to the well-known meteors weighing a ton or more that may be admired in various museums, but also to the cosmic dust that constantly sifts down through the atmosphere and, as it were, nourishes the earth with heavenly iron. W. Cloos, in an interesting work, has shown how these meteor showers radiate mainly from the constellation of Scorpio. Other observations indicate a connection between the forming of meteoric substance and the sun processes. Spectroscopic examination of the sun's corona has shown it to contain the same elements, qualitatively and quantitatively, as those in the meteorites. In discussing meteoric iron, therefore, we must keep in view the life of the cosmos, the sun sphere, and the constellations of the zodiac. The meteoric iron literally "precipitates" from these regions upon the earth.

THE PRINCIPAL IRON ORES.

With lead, the most important thing about its ore form was its combination with sulphur; with tin, the combination with oxygen. With iron there is no such preference. Its combinations with sulphur, oxygen, carbon dioxide, and water are all equally significant. We have four main groups of iron ores: sulphides, oxides, carbonates, hydrates. We must also mention iron arsenides, arsenic ores, and phosphates of iron. Iron chloride is present in the steam of volcanic exhalations. Wherever sulphur, oxygen, carbon dioxide, water, phosphoric acid, or arsenic acid display their activity, iron easily follows. This is true for the dead mineral kingdom, and much more so for the organic spheres.

Iron and Sulphur. Here the main ore is pyrite (FeS_2). It crystallizes regularly, mostly in cubes, but often in beautiful pentagonal dodecahedrons, called simply pyritohedrons. In their lustre, color, hardness, and weight these crystals show far more the nature of iron than of sulphur. The latter is conquered by the former.

We find pyrite in the oldest rocks, in ancient granite, in crystalline slate, in igneous rocks, permeating these with the tiniest crystals and following them into great depths. But it also concentrates into definite veins. Also, pyrite can be found in most sedimentary rocks, especially where masses of organic matter with their proteinaceous sulphur have encountered iron processes. Thus coal seams (as a residue of their old plant nature) and animal petrifactions contain pyrite as a mineral monument to the encounters of the sulphur and iron processes.

Pyrite always contains some gold. Frequently copper joins the company. The same combination is shown by markasite (also called spear pyrites), but this crystallizes in rhomboid form, represents a much younger formation that lies closer to the surface, and appears at lower temperatures than pyrite. Markasite belongs to a later "cooled-off" epoch.

Another important iron ore is magnetic pyrites (pyrrhotite). It contains iron and sulphur in approximately the same proportions, is of bronze-brown color, and crystallizes hexagonally. In contrast to ordinary pyrite, magnetic pyrites may be found in alkaline (low silicate content) igneous rocks, such as olivine or the serpentines. It usually contains cobalt and nickel.

Magnetic pyrites, the earthly form of ferrous sulphide, may be contrasted to troilite, a cosmic form that is composed in almost the same way, but crystallizes in the regular crystalline system. Troilite is found in meteoric iron, where it has cobalt and nickel for neighbors. When we consider that the meteor stones contain olivine, so that we receive from the cosmos a rock formation which in its terrestrial form contains magnetic pyrite, cobalt, and nickel, we begin to surmise mysterious interrelationships. Olivine is a silicate of iron and magnesium. All three constituents have a special relation to light. As to the plant world, we can say that the cosmos, whose power of light reaches right into the plant, ends in organs (the chlorophyll-bearing leaves) in which we meet iron, magnesium, and silica. Like dead counterparts of this living process, these same substances come flashing down from the same cosmos. We know from the investigations of Rudolf Steiner that long ago the entire mineral world was extruded from the cosmos as a dead nucleus of the earth.* In the properties of what is now dead matter we find concealed hints of living pre-forms out of which this matter has died.

The relations of iron sulphide to heat are interesting. At a higher tem-

---

* If we take these views as working hypotheses, an amazing number of phenomena become comprehensible.

perature, magnetic pyrites is formed; at a somewhat lower, pyrite; and at a still lower temperature, markasite. The first two point to warmer conditions of the earth, the latter to colder ones.

In iron-sulphur combinations we are confronted on the whole with ancient earth processes. In living beings strong mineralization is always a sign of age; with death, an organism enters wholly into the mineral kingdom. The more we trace it back into its youthful conditions the more we find it living, plastic, warm, unmineralized. The whole earth in primeval times was warmer, more plastic, more vital, more filled with proliferating life. This is shown by the coal forests and the giant forms of extinct animals. What today is part of the mineral earth, in the form of coal and limestone mountains, was formerly a part of life. In one substance, water, we can see to this day that it is a living substance of tremendous abundance, since two-thirds of the bodies of animals and four-fifths of those of plants are watery life juices. Many sea creatures can be said to be merely animated water. Life is the more encompassing form of existence, death the more limited one. To this day, water is always ready to dissolve into life, but never to depart therefrom as long as life is present. Only in serious pathological conditions, such as dropsy, do the dead physical forces take hold of the water within a still living organism. In the inorganic sphere water exhibits properties that are abnormal in the highest degree. These properties simply do not "fit" into the inorganic laws prevailing in the normal world of matter, but they fit well into organic existence. The observation of such scientific facts can open an understanding for anthroposophic research, which asserts that at the beginning of our earth there were organic, living processes out of which the dead mineral has only gradually taken shape. The superabundant, heat-filled, but lower life of those days has slowly toned down in order to make room for higher and more conscious creatures. In connection with these devitalizing events, sulphur was precipitated out of its organic protein combination into mineral form, with the help of certain metallic processes. In the case of iron it has combined with the earth's depths first as magnetic pyrites, later as pyrite, and finally as markasite.

The iron sulphides can preserve this ancient form in their deposits only if they are secluded from present conditions, enclosed and protected by an armor of rock. Wherever they come to the surface they immediately fall prey to the influence of the present earth, the forces of the atmosphere, which can only destroy their ancient form of existence. Air and water turn such iron ores to rust and salts; sulphates, hydrates, etc., arise; the so-called "iron hat" or gossan is formed. Oxygen forces the sulphur out of its combinations with iron. Thus the present earth nature clashes with the old.

Iron and Oxygen. In relation to the history of the earth, iron speaks its own language. Every important iron ore is connected with a different stratum laid down in the earth's past. Like the sulphur-iron compounds, the iron-oxygen compounds appear in three forms, pointing to three different ages of the earth.

Magnetite is the oldest of these forms. It is an oxide of iron, $Fe(FeO_2)_2$. In connection with oxygen, iron occurs in two phases of chemical valency: bi-valent as ferrous iron, and tri-valent as ferric iron. In each of these compound forms it shows a different character. At a lower degree of oxidation, iron is alkaline; at a higher degree, it is acid. The alkaline and the acid can combine with each other, and thus magnetite is a salt of ferric acid. It equalizes within itself the chemical polarities of alkaline and acid. In magnetite, what later separates into bi-valent and tri-valent iron remains as yet undivided. Magnetite, like the corresponding magnetic pyrites in the realm of sulphur, is magnetic. It has the ability to form beautifully clear octahedrons. By a subtle distribution of minuscule crystals, it imparts the dark color to most of the highly alkaline igneous rocks such as certain serpentines, basalts, etc. It accompanies the granite-splitting tin veins of Burma and Malaya (an interesting connection with tin), and it masses into whole mountains of iron in the mighty porphyry outcroppings of Kirunavara in Swedish Lapland. The high north (Sweden, Norway) and the Urals (Magnitogorsk) are the areas of such magnetite deposits, all belonging to old formations. Artificially, ferrous-ferric oxide is formed in hammer-scale in the forging of iron.

The next form of iron-oxygen compounds is ferrous oxide (FeO), although this never emerges as an independent mineral. But it is found, together with magnesium and silicic acid, in all old green schists, in olivine, diallage, etc. It lends these rocks their characteristic green color. In these, too, we have before us ancient regions of the earth, containing "light elements" * such as magnesium, silica, and iron, and showing in asbestos

---

* The relation of magnesium and iron to the light processes is fully disclosed in plant

something like a mineral counterpart of the plant. In these rocks iron is present as an accessory, merely lending the green color. As an extractable mineral or important ore it is not to be found there, just as in the plant it merely makes possible the formation of chlorophyll but does not enter as an ingredient. Here we will only mention that the schist formations were the mineral counterparts to the mighty plant-forming process, which was important in evolution after the sun separated from the earth and began to influence it from without (R. Steiner, *An Outline of Occult Science*).

In contrast, the tri-valent ferric oxide belongs to younger periods of the earth. While the bi-valent ferrous form lent a green color to all that it touched, the tri-valent iron produces red. Beginning with the Permian period it occurs mainly in the new red sandstones, then in the variegated sandstones (Keuper), that is, in the transition from the earth's antiquity to its Middle Ages, geologically speaking. When crystallized, ferric oxide (Fe_2O_3) appears as specular iron, hematite, in hexagonal-rhomboid form, but more often as thin flat scales sometimes forming "Iron Roses," or it penetrates certain primitive rocks giving a mica-like sheen; for the most part these are coarse-grained masses, another form of red hematite. Just as some silica no longer forms the well-shaped rock crystals but appears only as amorphous agate, so iron here occurs in coarse amorphous forms looking like red iron jasper, etc., instead of the clearcut shapes of magnetite crystals. The fine-grained, dense varieties are called hematite, or blood stone. Here we may mention the large deposits of red iron ore, as the coarse hematite is also called, found in Elba, Spain, England, New Zealand, near Lake Superior in North America, and in Germany. Thus we see that tri-valent iron can well appear as an independent mineral.

Iron and Carbon Dioxide. Iron has important connections with carbon dioxide as well as with oxygen, for another important ore is iron carbonate, siderite, or spathic iron ($FeCO_3$). In this mineral, iron shows its affinity to lime, for siderite is similar to calcite in form. The two minerals closely resemble each other; in fact, in the form of mixed crystals they sometimes present a combined mineral called ankerite. In siderite, iron is bi-valent, as

physiology. But silica likewise promotes the workings of light in the plant, as shown by L. Kolisko in her growth experiments, through the addition of silica at regularly decreasing intensities of light.

is calcium. Like calcium carbonate (which is calcite), iron carbonate dissolves in carbonic water to form bicarbonate of iron. This creates the so-called chalybeate springs (ferruginous water) and the chalybeate mineral waters. Iron, as well as lime, is thus drawn into the workings of the seasons, since the colder weather in winter and spring gives a higher carbon dioxide content to the flowing waters and thus brings more iron and lime into solution. In summer, however, when the water is warmer, part of the carbon dioxide escapes into the air, the bicarbonate of iron changes into the much less soluble neutral carbonate of iron and this appears as an insoluble precipitate; the lime does exactly the same thing. Thus we see how, in regions having distinct seasons, i.e., in the temperate zones, iron is brought into motion by the play of the seasons, and thus takes part in the rhythm of the earth. Relationships to carbon dioxide and lime manifest themselves here in the inorganic, but we grasp them fully only when the same processes are followed up in the human organism. For example, in the embryo the blood and the bone tissue both develop out of the extraembryonal mesenchyme. Or consider how the iron in the blood carries the carbon dioxide into the outbreathing, but a little is retained in the organism and, under the influences of the forces of the head organization, is brought together with the calcium process, which cooperates in forming the bones. But bone marrow is where the red corpuscles are born. Thus we discover, in the relationships of blood to carbon dioxide and lime (and also to oxygen), capacities of the iron process that reach their full import only in man. In the inorganic we see merely their prelude.

Many great iron ore deposits are pure siderite. We have already mentioned the Iron Mountain in Styria. There calcium also takes on a special form as flos ferri (a form of aragonite).

Hydrated Ores. The formative forces active in the present stages of evolution correspond to the destruction rather than to the upbuilding of the above mineral forms. All of these iron minerals, on emerging into the outer world, are immediately seized and transformed by air and water. An entirely different type of ore wants to arise out of this process, one that absorbs the *water element* in various degrees.

One such type is limonite (brown iron ore), $Fe_2O_3 \cdot H_2O$. It is the most important German ore. It is often dense, coarse, fibrous, spherical, kidney-shaped, and is also known as pea ore, bog iron, iron ochre, etc. Another such hydrated iron is called Goethite ($Fe_2O_3 \cdot H_2O$) or, according to its appearance, ruby mica, needle iron ore, etc. It is a happy accident that just this interesting and beautiful mineral should bear the name of Goethe, for it is a true product of metamorphosis. Something ancient proves capable of being transformed by the eternally creative forces of nature, and now manifests the newer forces of the earth. With these hydrated ores, in certain parts of the world, we can to this day see nature working at the formation of ore deposits. On the flat shores of certain Finnish lakes we find iron-bearing waters that have been formed by the leaching of weathered pyrite deposits in the vicinity. These waters flow into the aerated lake and precipitate iron hydrates, which accumulate in a few decades to such an extent as to justify periodic dredging.

* * *

The enumeration of ores significant for mining is now complete. The remaining minerals are much rarer, but we shall mention them because they reveal interesting aspects of the iron process.

Arsenopyrite (mispickel) is iron arsenide sulphide, a brittle substance of yellowish, silvery, metallic appearance. Though playing a minor role as an iron ore, it is the most important arsenic ore. Iron subdues not only sulphur but also arsenic in nature. This ability of iron will be properly appreciated when we consider the highly poisonous quality of arsenic. Arsenopyrite is frequently found in lead ore deposits; it gravitates towards the "precious" quartz formation (so-called because of its high content of silver glance). It occurs in association with the cobalt-silver veins of certain pitchblende deposits, and also some tin ore deposits. But it also contains gold. The world's largest arsenic deposit, at Boliden in Sweden, is at the same time a productive gold mine.

The veins that contain arsenopyrite (FeAsS) or the related Löllingite ($FeAs_2$) were formed under ancient conditions. Where they have been exposed, the arsenic has been reduced to arsenic acid by the oxygen of the atmosphere. Here again iron performs a good deed for living nature in that it binds this poison into an insoluble mineral, scorodite. This blue-green, black-green, or indigo-blue mineral, with its dull arsenical green streak, occasionally crystallizes but mostly appears in stalks or fibres, in grape form, or as an earthy covering. Rudolf Steiner found in arsenate of iron an important remedy for acute poliomyelitis. The famous mineral springs of Levico and Roncegno in Southern Tyrol contain an especially harmonious combination of arsenite, iron, and copper. Arsenic, tamed by iron, has become an important natural remedy.

Lastly, we will mention vivianite, a hydrated phosphoric iron, green to deep blue, rarely crystallized, mostly earthy. It is formed when phosphoric acid compounds encounter iron phosphate solutions. Therefore, it is frequently found in fossilized bones, where the calcium phosphate sometimes changes entirely into vivianite.

IRON AS A METAL.

Pure iron is not found in nature. It must be produced artificially. Then it becomes a soft silver-white metal of low strength (hardness 4.5), which can be drawn and rolled into the finest wire and sheeting. It is malleable and can easily be welded. This iron sheeting is as limp as paper. It can be magnetized, but loses its magnetism as soon as the inducing magnet is removed. Of itself, it is unable to retain the magnetic condition. It is a poor conductor of electricity, in this respect resembling lead and tin. It is also a substance much contracted within itself, having a small "atomic volume" (the quotient of atomic weight divided by density), and here it is close to carbon, whose affinity to iron we have already noted in other respects.

But the addition of small quantities of other elements will modify the properties of iron almost infinitely, and the development of modern technology rests largely upon skill in making steel alloys. Additions of carbon, silica, or the strongly formative hard metals with high melting points (chromium, tungsten, molybdenum, uranium, vanadium), make iron extremely hard, elastic, and strong, chemically resistant, almost precious, and permanently magnetic. Nickel and cobalt lend it strength and flexibility. These alloying techniques enable us to make tools that retain their keenness and hardness even when red hot.

Thus iron has the ability to absorb and retain the formative forces.

Iron, furthermore, is the metal that can best be filled and saturated with magnetism. It acquires for itself something of the magnetism that shows the earth's relation to the cosmos (magnetic storms, sun-spot activity, eruptions of solar protuberances). There is thus a connection between the darkening sun-spot activity and earthly magnetism.

The Chemistry of Iron.

In its connections with oxygen iron shows its nature better than in its ores or in its chemical behaviour. It is sometimes bi-valent and sometimes tri-valent, forming the oxygen-poor ferrous oxide compounds in which it is bi-valent, as well as the oxygen-rich ferric oxide in which it is tri-valent. There are many metals with two valences, but they usually show a strong preference for one or the other. Iron maintains a marvellous equilibrium between the two, an equilibrium highly significant in the processes of nature. It shifts easily from one to the other, and the most diverse events bring about this movement back and forth. It forms ferrous just as easily as ferric compounds, giving off and taking on oxygen with equal facility. It is, as it were, *the breather among the metals.*

Light is a powerful agent for converting ferric compounds into the oxygen-poorer ferrous compounds. Ferrous iron, which arises through the assimilation of the cosmic light element, can be called the more cosmic form and ferric iron the more earthly. A photocopying process (blue prints) is based on the light sensitivity of certain ferric salts. Iron is a real light catalyst. Traces of iron compounds can transform light forces into chemical reactions. This is the basis of a chemical process in Eder's photometer that measures the intensity of light. No wonder that iron helps the plant to form its light-organ, the green chlorophyll.

In contrast to most heavy metals, iron is not only non-toxic, but it can even counteract the effects of strong mineral poisons. We have already mentioned that in arsenical pyrites most of the arsenic in nature is fixed by iron and thereby rendered harmless. An important antidote against arsenic poisoning, the "antidotum arsenici," contains freshly precipitated hydroxide of iron as a main ingredient. Another important process showing the detoxifying role of iron in a wonderful light must be described here. By means of weathering, a great variety of metals are dissolved and washed down into the oceans, where in the course of time they accumulate to such an extent that the dissolved lead, copper, arsenic, and quicksilver would normally make life in the sea impossible. (Goldschmidt estimates the quantities at 58 grams of copper, 12 of lead, 2.9 of arsenic, and 0.3 of quicksilver per cubic meter of water.) But the sediments carried by the rivers into the sea always contain hydroxide of iron, which combines with these dangerous metals and precipitates them to the ocean floor as mud. Iron also counteracts the cyanogen process by transforming it into the non-poisonous potassium ferrocyanide. Hence the systems used to purify illuminating gas always contain iron hydroxide. Cyanogen is a terrible poison for the breath; iron, the breather among the metals, subdues it.

Only against carbon monoxide gas, arising from incomplete combustion of carbon, does iron prove powerless. The poisonous gas robs it of its metallic nature and changes it to a volatile yellow liquid, iron carbonyl.

Iron is largely soluble in its salts, and here it points to its relationship to liquids, to water. Iron contained in the rocks and soil enters the springs and rivers, which carry it out to sea. There, as described, it precipitates the other heavy metals as red deep-sea mud. There the reducing processes of organic sediments partly free it in the form of bi-valent ferrous iron, which then gradually rises upward again, combining with oxygen when it comes to the surface, hydrolizing as ferric iron, and sinking back. A circulatory system thus streams up and down in the oceans.

If iron in its salts were colorless and only bi-valent, we would look upon it as a kind of zinc. If it were only tri-valent and colorless, we would take it as something related to aluminum. Thus we see how color, the sensitivity to light and air, the combining with and dissociating from oxygen, are all a part of the nature of iron.

IRON IN THE REALM OF THE LIVING.

Green chlorophyll and red hemoglobin are two substances in the animate world that clearly manifest the principles of polarity and intensification discovered (or first enunciated in the full consciousness of their significance) by Goethe.* Iron is bound up with this polarity and intensification. It has to do with the polarity between plant and animal life, as well as with the intensification, the advance, from the vegetative life

---

* Goethe, *Theory of Color* and *Theory of Metamorphosis*.

processes that are etherically determined to the animal forms that are permeated by the astral. This refers not only to the contrast between the plant and animal worlds, but also to the transition from one to the other and the cooperation in animal and man, between the vegetative, assimilative, building-up processes and the soul-governed, degenerative, consciousness-awakening processes.

Iron must, as it were, stand in as godmother so that the plant may, "with and for the light," build the chlorophyll through which the cosmos shines into the plant and the plant, in form as well as substance, shines out into the world. As mentioned before, however, it is not iron, but magnesium, that enters the chlorophyll. The plant cannot, by embodying iron, form anything like hemoglobin, aside from a few abnormal cases where there is already a touch of the animal-like. (In the presence of light, certain papilionaceous flowers produce traces of hemoglobin in their root bulbs, which are filled with nitrogen-bacteria.) For over half a century scientists have been puzzled to see that chlorophyll and hemoglobin are so completely attuned to one another that they maintain the air in perfect equilibrium between oxygen and carbon dioxide (whatever the plant takes from the air is given back by the animal, and vice versa), and that the two breathing substances are so similarly constructed.

When chlorophyll is deprived of its magnesium and hemoglobin of its iron, we come in both cases to the same kind of substance, the so-called porphyrines. These have a strange quality. When injected into animals that are kept in the dark, they are harmless, but they turn into terrible poisons the moment these animals are brought into the light. The light becomes a lethal power for animals so treated. But the addition of metal, whether magnesium, iron, or copper, takes the poison out of the porphyrines and transforms the light to an element that builds up life. The porphyrines may be separated into four pyrrole rings; these are compounds, combining in ring-like form one particle of nitrogen and four of carbon. But these pyrroles point toward important amino acids that arise in the disintegration of protein and contain similar ring formations in which nitrogen and carbon are directly associated. Such amino acids are tryptophane, proline, oxyproline. The nearest relative to pyrrole is pyrrolidine. Here we return from dead formulas to something alive and visible in the

Chlorophyll a   Hämatin

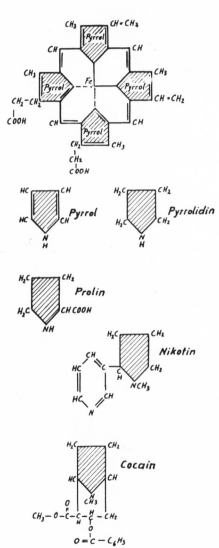

Porphin

Pyrrol

Pyrrolidin

Prolin

Tryptophan

Nikotin

Atropin

Cocain

76

plant world for we find pyrrolidine in many plant poisons, such as the al-
kaloids of tobacco (nicotine), of henbane (hyoscyamine), or belladonna
(atropine), and of the coca bush (cocaine). Thus we see that if we kill the
living protein and allow it to decompose in a certain way, we can break it
down into the above amino acids. In the alkaloid plants, the poison organ-
izes itself out of the living protein. Does this not suggest a mortifying
process that takes hold of the alkaloid plants, destroys part of their protein,
and leads to the formation of poisons? Indeed, is not this process, in a
milder form, present in every plant, though only where it produces its
chlorophyll? Does not a similar process take place in the blood, the alkaloid formation representing merely an excess? The metals are, in consequence, a natural remedy and antidote against something which, though
necessary to life, nevertheless appears as a life-destroying principle.

Let us return once more to the alkaloid plants and the forming of poisons in the plant world. This is something abnormal. How does it arise
among the plants? This can be understood only when we view the abnormal as a one-sided exaggeration of the normal. For the plant can develop
only processes that conform to its own nature.

Consider the transition from the forming of the green leaf to that of the
blossom in normal non-poisonous plants. The green leaf is the primal
organ, the clearest expression of the plant. In rhythmical recurrence, from
node to node, it repeats itself over and over again. But soon a counter-
process sets in against this continuously sprouting life. The sprouting
stops, the leaves contract, the nodes join with each other until they melt
into a formation that, though still green, opens into a chalice, manifesting
the plant in a new way, on a new level. A higher organ arises, which has
lost the green color and the life-building faculty of the leaf but, in the
more perfect form of the blossom, makes the plant nature directly evident,
condensing it to a visible picture. By their blossoms we know the plants.
According to Goethe (whose theory of metamorphosis, as expanded by anthroposophy, we are following here) the plant is a "sensible-supersensible"
being. The sensible, the physical body, is joined by a supersensible member, the body of formative forces. In the green continually sprouting forms
of life, in the leaves, these formative forces (the etheric body) come to expression as an active field of weaving life forces. In the blossom, however,
the reflection of a higher member appears weaving about it, directing it

from without, and causing at first an abatement of the etheric processes, until it touches the plant from without, as it were, in the flower, and reveals itself in the coloring. This higher member we call the astrality of the plant. It is related to the animal nature, and with it the plant lifts itself to the border of the animal world, whence come the innumerable connections between blossoms and animals, particularly insects. (Many blossoms, for example, are negative counterparts of the bodies of the insects that visit them). The great distinction between animal and plant is that the astral principle of the physical plant always remains outside it, irradiating it from the spiritual-cosmic expanses; whereas the animal has taken the astrality into its body, has incorporated it, and now directs from within a multitude of activities and processes that the plant receives from without, from the cosmos. Thus the plant really has only one organ, the leaf; it is an organ wholly without inwardness, a mere surface for the light to touch, actually a kind of hollow sphere whose center is the sun. The animal, however, has its own organ-cosmos, genuine inner and hollow organs that have their centers within themselves in every respect.

In the flowering process there occurs a delicate contact between the life sphere of the plant and the higher astral sphere. But this contact paralyzes the vitality, brings growth to an end, shrinks the leaves, eliminates the green color. It causes strong symptoms of degeneration in the metabolism of the blossom; in heavily flowering plants, this may go so far as to form decaying matter. In narcissi, arum plants, and some varieties of lily, we find traces of indol, skatol, and related substances usually associated with putrefying protein.

If this contact of the etheric and astral spheres becomes more intense, or if the astral sphere penetrates deeply into the life structure of the plant, the breakdown and paralysis of the life processes become abnormally strong. In the hyoscyamus plant (henbane) or the mandrake root (mandragora), the blossoming process invades the sphere of the leaves and sprouts abnormally and prematurely; it fairly presses into it. Here we see how a plant, having barely reached the stage of leafing, is overwhelmed, as it were, by the blossoming process. A similar observation may be made with belladonna, stramonium (jimson weed), and many other alkaloid plants.

If we understood the reversal of the relationship between the processes conditioned by the astral in the animal and the etheric processes in the plant, the inversion of the relation between "without" and "within," we will find nothing strange in the reversal of the breathing process into its completely obverse mirror image. The contrast and the kinship between the plant, which inhales carbon dioxide and exhales oxygen, and the animal, which inhales oxygen and exhales carbon dioxide, corresponds to the resemblance and contrast between chlorophyll and hemoglobin, as well as to the role played by iron in plant and animal. To lay hold upon iron directly and make use of it in an inner breathing process is possible only for a creature in which an astral body is active.

In the human organism, too, the etheric body works as a constructive force, vegetatively. Opposing it destructively is the work of the astral body. This occurs particularly in the activity of the nervous system, which is a system of organs for the impulses of the astral body. The result is that the life processes are blunted and the death forces are taken into the body. But these breakdown processes are important for man, being the bodily foundation for the appearance of consciousness. Only a being with an astral body can have consciousness. Only a being with an ego can have self-consciousness. But the shadow side of these processes is breakdown and illness. Something like a "normal sickness" pours continually from our consciousness pole into our bodily organization.

This "sickening," as Rudolf Steiner frequently explained, reaches as far as the blood. But he also pointed out that in the blood a healing process arises that is the real prototype of all healing. The meeting of these two processes (the astral-conditioned one that causes a "sickening," and the ego-controlled healing one) is imaged in the meeting of the poisonous porphyrines with iron, which here again, in the organic world, draws the sting from the poisons. The porphyrines have a basic connection with poisons, as shown in their relations to the strong plant toxins. It is thanks to iron that we can bear the destroying illness-evoking effects of our consciousness.

A TYPICAL IRON PLANT: *URTICA DIOECA.*

Because of its iron content, and even more because of its special way with iron, the large stinging nettle, *Urtica dioeca*, is an important medicinal herb.

The normal plant, with its root, leaf, and flower, is distinctly a threefold creation. By contrast the nettle seems abnormal. The leaf predominates, the flowers and roots are insignificant. Leaf-pair is piled on leaf-pair; one node brings forth the next; each repeats, with two side-leaflets, the germinal leaflets of the first beginnings of growth before releasing its main leaves. The plant enjoys repeating itself without ever arriving at a decisive metamorphosis. It is actually a rhythmical leaf-stalk throughout. This even continues underground as a creeping root stock from which the small rootlets issue. But every surface node is equally ready to produce such rootlets at any opportunity; for example, when the plant is pressed to the ground and remains for a time in contact with the earth. Toward the upper end, this leaf-stalk sends up the unpretentious greenish clusters of blossoms, which from a distance might be taken for feathered leaves rather than flowers. Leaf and stem are thus the predominant and decisive elements of this plant. For the nettle the meeting of water and air that occurs continually in the assimilating leaf and in the stem (which always stays green) is an important event, even when it takes place outside the plant. The leaves are exceptionally sensitive to the interplay of light and shade, dampness and dryness, in their environment. The edges change their form, from round indentations to sharp sawteeth, according to whether they grow in moist darkness or bright dryness. The nettle makes its home at the forest edge, in clearings, near fences, along the roadside, on the river bank, wherever there is a living interplay between light and darkness, water and air. This interplay reflects a unique cooperation of the etheric and the astral, since the former prefers all liquid processes, and the latter those connected with the air. This cooperation takes place in a special way in the *periphery* of the nettle. The vital element pushes through the leaf-surface and extends the growth beyond itself into a strong hair formation. This then rigidifies through intensive mineralising. The hairs are gland-like cells that grow into hollow needles, which contain calcium at the bottom and silica towards the point. They break easily and spill their liquid content outward. This content is the product of an abnormal protein decomposition. It produces an inflammatory toxic protein, which is tinged with formic acid and is said to have a certain resemblance to snake poison, and above all to bee poison. It also contains histamine, an especially interesting product of the breakdown of protein. Histamine is likewise present in bee-poison, and is thought to be the cause of the swellings resulting from bee stings and nettle burns. Closely related to histamine is histidine, an amino acid found in hemoglobin, in the proteinaceous component of the red corpuscles. Histamine is a substance that can absorb and transmit certain impulses of the astral body. It stimulates, for example, the secretion of the gastric juices; it is regarded as the carrier of the pain sensations in the nerves. Also, the juice of nettles contains "secretins," substances that arouse the activity of the pancreas, thus extending their effects into organs that appear only in beings that have astral bodies, i.e., animals and men.

These phenomena, taken all together, show that the nettle is enveloped by a peripheral toxic process that breaks down the protein at the periphery and converts it into poison. (Nettle protein disintegrates easily; an extract of the juice is quick to rot and stink). The resultant substances reveal that at the outermost border of the nettle its etheric configuration is seized by astral forces, which induce the nettle to form poisons as described. In certain tropical varieties, the poisons have such a sting that careless contact may cause a seven-day fever.

But why does the nettle never turn into a strong poisonous plant in the usual sense? Obviously a counter-process prevents the forces active in the periphery from entering into the plant. This counter-process lies, it seems to us, in the nettle's ability to permeate itself with iron to a special degree. *Urtica dioeca* contains a remarkable quantity of iron. But it is also a conspicuously *green* plant; it forms an inordinate amount of chlorophyll, so that most commercial chlorophyll is derived from it. *Urtica dioeca* contains over 6% iron oxide in its ash. In addition to chlorophyll the leaves contain a great deal of carotene, which is a bright yellow pigment that is intimately connected with light. In man, this changes to Vitamin A. This strong relation of the nettle to the iron process extends into its environment. As Rudolf Steiner points out, it regulates the influence of the soil iron on the plants, neutralizing the harmful effects where there is too much iron and aiding the plants where there is too little.

To sum up, the protein of the nettle decomposes and decays readily. This means that the plant offers easy access to the peripheral astralizing processes whose substance-destroying forces make the plant poisonous. But, by virtue of its high iron content and its vigorous iron process, it heals the decaying tendencies and keeps the astralizing processes away from its etheric processes. Nevertheless it forms substances that, in man, stimulate the digestive glands to increased activity, something that is normally caused by the astral body taking a more powerful grip upon the etheric. The remarkable dynamics of the nettle enable the substances derived from it to stir up such increased action. Furthermore, these substances stimulate the formation of blood. About the fourteenth year, the human astral body must penetrate the physical-etheric of the child's body in greater measure than before, thus bringing puberty about. This is also the age when anemia is most likely to arise. The urtica-iron dynamics induce a healthy connection of the astral with the heretofore more vegetative bodily activity. Since the astral body uses the aeriform organization as its particular instrument, the aerated blood with its iron-bearing hemoglobin is, because of its iron content, an important regulator of the measure of this penetration. By increasing the blood-iron we place the soul-spiritual nature of man more energetically upon the physical ground. The dreaminess of adolescence changes into a vigorous grasp of life.

We have termed iron the breather among the metals. Its behavior in the inorganic justifies this designation, but the term is even more apt when we contemplate the breathing of plants, animals, and man. In man, we might even speak of a threefold breathing, with which the iron processes are closely connected. The following sections will expound this.

LIGHT-BREATHING.

The most important pigments produced in all three kingdoms of nature depend upon iron, as was mentioned before. In man, too, iron cooperates in this work. Not only the peach blossom color of the skin is due to it, but also the forming of melanin, the pigment that protects the eyes and skin from too much light. Melanin arises in the presence of iron (and copper) although, like chlorophyll, it actually contains no iron in itself. Wherever this pigment is lacking, as in albinos, there is hypersensitivity to light and the eyes and skin are easily inflamed. Such an organism must keep away from light; it cannot transform and "digest" light. The "light breathing" of the skin is not properly regulated. Iron provides this regulation. Pyrite may be tried as a remedy against albinism.

Since iron as a substance stands in the "center" of the hemoglobin, it forms the basis of man's air breathing. But the subsequent conveyance of the inhaled oxygen towards the combustion processes in the metabolism also occurs—so it is assumed today—by way of the iron-bearing ferments (Warburg's breathing ferment). The breathing process expresses the relationship of iron to oxygen on the human level. Thus we must view the activity of the *lungs*, too, in connection with iron. In disturbances of the lung process, in lung diseases such as pneumonia or tuberculosis, various iron preparations may be the right remedies. Through iron, the breathing process is closely bound up with that of the blood formation. How iron reaches into the latter, thus becoming an instrument of the soul-spiritual in man, has already been indicated. In the hemoglobin iron confronts protein, taking hold of it as the latter decomposes and incorporating it healingly into the organism. There is a polarity between protein and iron processes. Consider, for example, lower animals such as the oyster, in which iron as a breathing element plays no great role; they are mere metabolic creatures, shapeless living protein, with only a trace of dull consciousness. Compare these with the higher animals that have "achieved" iron. This achievement means aeration, warming, shaping, structuring, inner skeletal formation, ensouling. Iron helps wake the proteinaceous body from its vegetative sleep and opens it to the forces of consciousness. The antagonist of iron in this region is sulphur, which stimulates the forces of vitality in the protein but dulls the consciousness throughout the organism.

Excessive metabolic sulphur processes may be combated with iron. It not only strengthens the soul's grasp of the body, but helps the ego to permeate the body with its impulses. For iron helps with the proper heating of the body, and in this heating the ego can make its impulses felt. In the hemoglobin iron is an almost mineral substance and can be crystallized. Therefore it infuses the blood with mineral earthly laws, but these are governed by the ego. This process enables the ego to assert itself forcefully in the physical world. (An anemic person, lacking sufficient iron, is unable to heat himself properly and suffers from lack of resolution; he is not only poor in blood but weak in his ego and unequal to his earthly tasks.)

Liquid Breathing.

A special metamorphosis of the blood process in the metabolic realm is the bile process. We had to seek the "light breathing" in the sense-nerve organization and the "air breathing" in the rhythmic organization. Now we come to the metabolism.

The biliary pigments are an important constituent of the gall. They arise out of the hemoglobin as the iron departs from it. The bile fluid now performs a movement similar to exhaling and inhaling. It streams into the intestinal tract, combines with the fats in the food, makes them soluble, and is taken back with these into the interior. In the gall we have in a certain sense a transformation of the blood. In this new form it ventures out of the innermost recesses of the body and, with its strong converting forces, confronts the stream of nutriment, herding home what it is able to transform. We may call this "bringing home the bacon." In congestions of the bile-flow, an iron therapy such as massage with a ferrous salve can be effective.

The Cultural History of Iron.

We have considered iron as pure matter and we have investigated its role in the organic realm. We will see it in a new light when we follow it from nature into civilization, into the works of man.

In the ancient mythological phase of human history, the smith is a lordly figure. He presages the secrets of the Iron Age. This is the time in which to take hold of iron externally and internally. Externally, the smith works with fire, bellows, and hammer, a son of Prometheus shaping iron into tools that enable him to master the physical plane. Internally, we kindle the blood's fire with our breath; with every pulse-beat we develop the thrust of the will in our organism and let this will stream into our active limbs. What is happening here except that the "inner smith," latent in all of us, is bringing forth the iron-forming smith as his external counterpart? Internally we work on the blood-iron, externally on the forging iron. Tools and weapons come into existence here. Man needs these when he no longer experiences himself as a child nourished at the breast of nature, sheltered and carried by the surrounding world, but is forced to win his living in defiance of nature and to defend himself against the enemies that oppose him on his self-chosen path after the Golden and Silver Ages have run their course. This relation to iron has been portrayed as a spiritual secret of human evolution in fairy tales, such as the "Iron Hans" of German folklore, and in the Finnish epic, *The Kalevala.*

But the Iron Age in the full sense of the word begins in what we call modern times, because now, in contrast to the Middle Ages, man is developing and asserting new faculties of soul. These are faculties of a heightened wakefulness of the soul. They arise because for man all of nature is, as it were, dying out, so that only the material side of existence remains accessible for him. For the physical side of nature, which is all that the waking senses of the daytime can experience, is what has already become and been completed, the "natura naturata," the death-side of existence that is turned toward the senses. "Natura naturans," the unfinished, living creative world, lies beyond the senses. It was lost to us with the decline of the Golden and Silver Ages. What can be grasped by counting, weighing, and measuring is dead.

But in this world of lifelessness man can feel himself to be free, free as never before. Forces of self-consciousness awaken in him that were unknown in any earlier times. A new limb of the soul grows up in him that may justly be called the "consciousness soul." It exercises itself by searching for the spiritual laws that work in the dead world of the senses. Thus it becomes the lord of outer nature. The true Iron Age, led mainly by the peoples of the West, arises as a creation of iron and steel. Iron is harnessed to purely material goals. The shackles by which man has enslaved nature are forged of iron. In quantities millions of times larger than ever before, iron is extracted from the earth. It is the stuff in which the engineer's abstract thought-forms incarnate. The world of technology carves its place out of the living body of nature. This means liberation from the natural forces, which are made to play against each other so cleverly that they cancel each other out. But this liberation rests upon the forces of the earth's death.

What is so exclusively based on death forces must bring death. It has become the destiny of iron, more than of any other metal, to serve war and destruction. To lift this curse from iron, must we not be able, in some place in the world, to overcome the death forces in iron?

Let our own nature teach us. In the crystallizable hemoglobin, iron carries the laws of mineral nature (which is dead) right into our blood. Here this mineral nature is overcome by our ego and pressed into its service. The iron in the blood no longer obeys the laws of nature, but the will impulses of the ego. Look at a human limb motion, by which we mean one that is consciously willed, freely decided. It is preceded by a movement of the blood, and therewith by a movement of iron, into the organ concerned.

Our modern consciousness, on the other hand, is tied to the activity of our senses and of the nerves and brain connected with them. This activity results in processes of decomposition, destruction, and death. By virtue of its iron content the blood sets up healing processes to oppose these pathological accompaniments of the unfolding of consciousness. Through iron the ego is connected with forces that can bring the processes of sickness and death into the service of human evolution. If we understand and experience the spiritual force that makes use of the iron within us, if we thereby take hold of a part of ourselves that otherwise acts only in our unconscious, then we gain the power to set goals for the outer Iron Age in which the human spirit can reveal its true nature. Goals that are illuminated by knowledge and suffused by the ego will some day be realised in such an approach to iron. This should be the fruit of the "Iron Age."

VI

GOLD

All the metals are wonderful strangers on earth, but gold most of all. Although the normal earthly processes tend to eliminate the metallic state of existence, to destroy its lustre, permanence, malleability, and firmness, through rust, weathering, oxidation, and calcification, they are powerless to do all this to gold. In the precious metals the metallic nature stands revealed in its highest perfection. The prototype of the precious metals has always been gold.

The non-precious metals succumb to the attacks of the earth forces and lose the imponderable side of their nature; light, heat, outraying energies, are released, either through rust (a slow form of burning) or through other combinations with earthly matter. Thus they lose part of the cosmic aspect of their nature. On the other hand, it is these imponderables that in the smelting furnaces must be added to the ores so that the pure metals may be derived from them. The individual metals are deemed to be more or less precious according to the extent to which they retain this cosmic aspect of their nature. Gold, the most precious metal, simply cannot lose its cosmic aspect.

For a true understanding of the essential nature of a metal we must know what kind of cosmic entity penetrates and constitutes it. We shall attempt to show that with gold it is a sun-like kind, that gold must properly be associated with the sun.

We shall begin by tracing the forms in which gold manifests in the earth. Thus we start with the mineral processes most accessible to our senses, and then ascend the kingdoms of nature until we finally come to man, where gold demonstrates its highest earthly form.

GOLD IN THE BODY OF THE EARTH.

Paradoxically, although gold is rare and precious, it is widely prevalent. There is no continent or climatic zone without considerable gold deposits. It appears everywhere, albeit in finest distribution. In the strata that can be chemically investigated with our present methods, it is present in the ninth decimal potency ($1:10^9$), or one part per 100,000,000. Seawater contains it in about the seventh decimal potency. Granite contains about one part in a million, diabase 0.7, basalt 0.2. The more silicic acid the rock contains, the richer it generally is in gold. This relation of silicic acid to gold will emerge even more clearly in the description of the gold deposits.

But only a small part of all this gold is recoverable. Almost all of it serves only to "homeopathize" the earth with gold. In few places does it occur in sufficient concentration to be worked.

But the previously mentioned fact should be kept in mind, that rich gold deposits are to be found on all continents. The world production in 1938, for example, was: North and Central America, 27%; South America, 4%; Africa, 40%; Asia, 19%; Australia, 6%; Europe, 4%. The yields since the discovery of America are estimated to be: North and Central America, 31%; South America, 5.5%; Africa, 31%; Asia, 4.7%; Australia and Oceania, 14%; Europe (including European Russia), 12%. A close look at the deposits shows that they frequently lie in uninhabited places, even in deserts. Africa, the continent with the largest deserts on earth, the lion continent, is at the same time the richest in gold. But Africa is also the continent that, in its climatic structure, most clearly shows the effects of the sun. The distribution of its flora is most nearly symmetrical. In the extreme north as well as the south, this continent is really Mediterranean. The flora of its southern tip and that of the Mediterranean coast have much in common. Touching the north of the Union of South Africa and the southern edge of the Mediterranean coast are broad desert belts. Each of these adjoins a prairie belt. These in turn, toward the middle of the continent, converge into a central belt of primeval forest in the equatorial zone. Thus this sun-continent is also the gold continent, the center of gravity for gold in the earth.

It may be stated further that those areas that became solid land early in geologic history show rich gold deposits. In addition to South Africa, this means Australia, India, Canada, and Scandinavia. Likewise, the mountains of the North and South American West coasts. In Europe the chief gold sources were the mountains of the Spanish northwest, the Alps, the Carpathians, the mountains of the Balkans, the Sudeten, the Harz, and the Urals. More about this later.

Gold in Silica and Iron. The "Old" Gold Formation. Extremely diverse regions of the earth, many different rock formations, shelter the physical concentrations of metals that represent commercially valuable deposits. This was pointed out in connection with lead, tin, and iron. A region of silica or of iron and sulphur opens before us when we look at the important "old" gold formations, the silicious gold quartz veins formed in primeval times. Silica is a substance that is connected everywhere in nature with certain qualities of light. In the mineral kingdom it forms the clear rock crystals and most of the precious stones. In the plant kingdom it facilitates the effects of light. In man it works in the sense organs, making the eye clear and translucent. Gold enters readily into company with quartz and colors it delicately, although because of its own aristrocratic nature it does not combine with it materially. Thus there are veins of pure gold quartz ore. In its liking for silica's relation to light and form, gold reveals its own kinship with the light.*

But we also find gold in the pyrite veins of crystalline primeval rocks. Here it shows its relation to iron and sulphur. Within these veins it penetrates to great depths, descending deeply in the direction of the earth's gravity, more deeply than most other metals. The world's deepest mine shaft runs down a gold-bearing pyrite vein in Minas Geraes, Brazil, to a depth of over 8,000 feet. Gold shows not only a kinship with light but also a relation to gravity, a polarity that will face us repeatedly as we proceed in our considerations. Gold as a metal between light and weight is a motif whose strains will resound for us again and again.

In addition to this polarity, a marvellous trinity should be noted be-

---

* Compare the growth experiments with silica by L. Kolisko, mentioned in Chapter 5.

tween silica, iron, and sulphur. The interweaving of these three emerges most perfectly in man. The silica processes are connected primarily with the sphere of the senses and the head; the iron processes, with their relations to breathing and blood, point to the middle rhythmic sphere; the sulphur processes are related to metabolism. It is man as a whole that stands before us when we pursue the dynamics of gold in the world of matter and seek for an image that expresses their inter-relationships most completely. Gold links itself harmoniously both to the clear crystal form-world of silicic acid and to the fiery-volatile world of sulphur, when the latter is tamed and governed by a third, the sphere of iron.

Iron sulphide can become especially rich in gold when it combines with arsenic, appearing as arsenical pyrites or arsenical iron sulphide. Arsenic is an element that, in cooling, changes immediately from gas to solid, leaping over the fluid condition. It contains strong tendencies toward solidification, which are apparent also in its hardness and brittleness, as well as in its ability to make hard and brittle any other metals with which it is alloyed. In the organic realm arsenic, which is a vehement poison, damps down the etheric activities and energizes the astral body, stimulating it to enter more strongly into the physical. It is, therefore, an important remedy. In arsenical pyrites it condenses the gold substance.

Gold found in the old formation contains, as does all natural gold, small quantities of silver, but in this formation gold predominates by far. The silver content is only minor in contrast to the formation about to be described.

Gold in Association with Silver, Copper, and Antimony. The "Young" Gold Formation. A second impulse toward the formation of gold deposits occurred in a much later epoch. To be mentioned here are the so-called "young" gold ore veins found in late eruptive rock formations close to the surface. These originated in places that were permeated by heat-processes longer than the old crystalline primal rocks that hardened first. Here, too, silicic acid appears as a companion, but in the form of hornstone, chalcedony, even opal. This leaves the silicic acid closer to the colloidal form in which it occurs in organic compounds, where it is not yet entrapped into crystalline form to the same extent as quartz. In these young veins we find an entirely different grouping of metal sulphides than in the old veins. Silver, mercury, copper, and antimony occur as sulphur compounds, forming as fahl ores (tetrahedrites) and related minerals, the realm in which gold is found. In these veins silver outweighs gold up to a hundred to one. Only the much higher value of gold leads us to call them gold veins. Again we find the trinity of silica, metal, and sulphur, but in modified form. Iron is lacking. Here the gold, in the form of the tellurides, calaverite and sylvanite, combines directly with the sulphur-related tellurium and selenium.

In Europe, the Alps and the Bohemian mountains contain gold of the old formation, whereas the Carpathians have the young one. This is clearly mirrored in the composition of the placer gold in the Danube, which is much richer in gold and poorer in silver near Linz in Austria than it is in Hungary.

The occurrence of gold in two such different forms is a reflection of the two completely different phases of time during which gold was infused into the body of the earth.

EARTH HISTORY AND GOLD DEPOSITS.

The picture of evolution in Rudolf Steiner's works (for example *An Outline of Occult Science*) offers a key to the understanding of many riddles. We shall test this with our problem of gold.

For the spiritual-scientific view, the solar system is a mighty organism that has evolved to its present form from germinal, unorganized, life-force-permeated primal beginnings. In so doing it has gradually exuded the life-less minerals, as a snail secretes its shell or a tree its bark. (It is not feasible to enter upon the details here. Those interested may follow them up in the original literature. A comprehensive work on the subject is G. Wachsmuth's *The Evolution of the Earth. Cosmogony and Earth History—an Organic Growth*.) The "embryonic" condition of the earth, however, is to be thought of not only as sustained by life-forces but as permeated by soul-like nature and by spiritual beings. The aging of this gigantic organism is accomplished in steps of gradual condensation and materialization, leading to ever greater segmentation, to the separation of the sun, the planets, and the moon, and to the "secretion" of the nature kingdoms, etc.

This primeval sphere originally contained the sun, with its forces and beings. At a certain point in evolution the sun detached itself from the common body, which still embraced the earth and the moon, and began to work upon these from without. Certain of these sun-effects consisted of etheric streams, which led to the formation of gold. Out of a condition consisting merely of color and light, it condensed to air or vapor, then to liquid, and finally to the solid state of our present gold. In a profound sense gold belongs to the sun. Relationships between gold and the sun exist to this day. We have mentioned how L. Kolisko discovered impressive phenomena by following up certain reactions of gold solutions to other metallic salts, using the capillary-dynamic method in filter paper. She showed, for example, how the gold salt reaction took an entirely different course during an eclipse of the sun, was able to demonstrate the "sun-sensitivity" of gold.

At the time of the sun's separation, the earth with its beings had attained a plant-like stage. The dependence of the plant world upon the sun is a reflection of this. But the plant-like condition of that time was of an entirely different nature than the plant world today. It wove in the *atmosphere,* which was alive through and through, like a kind of fine volatile protein. Cloud-like, with green colorings appearing and disappearing, these plants floated in the proteinaceous atmosphere. This atmosphere was filled and energized with mighty processes of sulphur and silicic acid, whose dynamics conveyed to the plants the forming, shaping, etheric effects of the light. All substances were connected with the dynamics of life processes, as in the organic world of today. They had not yet developed the laws of the inorganic, the lifeless. The substances that we are familiar with today are only the petrified corpses of these dynamic forces. *Their original nature is living functions. Matter is only their end.* Taken into the life of animate beings, these substances even today are immediately converted into the dynamics of life. Silicic acid forms and shapes in plant and animal to this day, and transmits light activity. Sulphur still lives in the proteinaceous sphere of all animate beings.

Evolution proceeded toward further condensation. Out of the earth's biosphere the processes tending toward death were extruded and became the beginnings of the primeval rock formations, together with what was to become silica. The mineral world dropped out of the world of life. Into the condensation processes, however, and into all of earthly activity, there shone the sun and the planets, which at this time had also separated from the original common body. The metal-creating impulses thus radiated

92

from the universe into the earth. These impulses were of etheric nature, and they combined with the plant life. The resulting "secretion" was the mineral formations, into which the ore-veins wove themselves. The excess of sulphur in the atmosphere was absorbed by the metallic, giving rise to the pre-stages of our pyrites, glances, blendes, and metal sulphides. The earth was especially permeated with iron after the separation of the sun; for, with the expulsion of Mars and the continuing condensation, the original common body, (spatially still much larger than today's earth) contracted further. The earth (still including the moon) passed through the sphere of Mars and received from it the cosmic gift of iron. (In these descriptions we must, of course, not think of today's conditions or forms of matter.)

Thus the situation of our planetary system at that time was as follows. The earth, still containing the moon, confronts the sun, from which Mercury and Venus have not yet been separated. The earth has experienced the impact of iron through the influence of Mars. Out of the proliferating plant life, which is under the influence of the sun, the earth secretes, in the course of further condensation, the mineral world. Silica rock is formed and the metallic sulphides congeal into it. Within all these activities gold, the gift of the sun, condenses. The sphere of the "old" gold formation, the pyrites and ferrous sulphides, takes the gold into itself.

Evolution continued. The moon-forces in the earth worked too strongly toward condensing. The light of the sun was less and less able to penetrate the earth's covering. Darkness increased. In the end, the moon and its excessive hardening forces were expelled. Now the sun's influence could again penetrate the purified atmosphere, which had once more become transparent. The plant and animal worlds became more and more attached to the now liquid but continuously solidifying earth. The atmosphere slowly began to resemble that of today.

Meanwhile the sun had expelled Venus and Mercury. The moon, now working on the earth from without, gave the impulse for the formation of silver. Venus and Mercury gave the impulses for copper and quicksilver, and the cooperation of Moon, Mercury, and Venus produced, according to Rudolf Steiner, the impulse for the formation of antimony. Surveying the events of that period as a whole, we find what comes to expression as the second impulse for gold deposits, the "young" gold-silver formation, in which iron plays no role but copper, quicksilver, silver, and antimony together create a lodgement for gold.

FURTHER GOLD CONDENSATION PROCESSES.

Even the richest auriferous quartz-veins contain gold only in minute, hardly visible flecks. Values of three ounces to a ton (about one part in 10,000) are considered extraordinarily high. But a remarkable concentration process can bring the gold out of this tenuous distribution into masses sometimes weighing many pounds. This is the "cementation" process, which can be explained as follows.

Gold, as has been pointed out, stands like a rhythmically connecting principle between quartz and sulphur. Now we must point to a middle area, a heart area, in nature in which gold, also spatially, appears as a contraction of matter. We follow a vein of pyrite ore which, in its ferro-sulphuric nature, contains gold in the finest dilution. To begin with, this is a formation deep within the earth, originating from the oldest earth conditions. On coming to the surface, however, it is exposed to the influences and forces of the upper world as it is today. The atmosphere, above all oxygen and water vapor, seizes it at the point of emergence. The iron greedily breathes in the oxygen and thereby thoroughly transforms itself. It changes from its "bi-valent" condition as a sulphur compound to a trivalent oxidized one. The sulphur is burned up, it becomes acid; and out of these two processes arises ferric sulphate, tri-valent iron sulphate. The upper deposit turns to rust, resulting in what miners call "iron cap." With the dew, rain, etc., this iron salt seeps down in solution. But it has a remarkable ability, especially in the presence of chloride (which is widespread), to dissolve precious metals, among them gold. This gold salt solution likewise sinks down, and the whole process continues as long as pyrite continues to be accessible to the "upper world," the atmosphere. But now this oxygen-related sphere of tri-valent iron comes to grips with the "lower," the untransformed world of sulphur of the old bi-valent pyritic iron; the tri-valent ferric sulphate is reduced again to bi-valent iron salt, ferro-sulphate. But this is a powerful means of precipitating gold from its salt solutions! The flow of dissolved gold seeping down from the oxygen sphere into the sulphur sphere finds a position of balance, a kind of "dynamic diaphragm" in which it combines into nuggets. It finds its way into this area of balance by the power of iron. Above oxygen reigns, below, sulphur.

In this entire cementation process nature presents us with a picture of the gold process even in the inorganic mineral world. Extending into two polaric force spheres, it manifests visibly and tangibly in their center, finding its "place" dynamically as well as physically in the balance between both poles. In man, too, we shall find gold's seat of activity in the center, for it belongs to the sphere of the heart. In man we have on the one hand the upper region, open to the world of light and connected with the silica processes, the region that includes the breathing man; on the other, the lower metabolic region, which deals with the forces of gravity, uses sulphur to build and rebuild substances, and leads over to the iron-permeated blood circulation. Between these two poles, in the center of the rhythmic system that serves them both, stands the heart, which owes fealty to gold. We shall return to this when discussing the healing power of gold.

GOLD AS A METAL.

Gold has a tremendous "intensity of being life." This is shown by phenomena of light and color. The serene splendor, the shining magnificence of the impression that the sight of gold makes upon our minds and senses is the highest enhancement of which the brightest of all colors, yellow, is capable. Yellow has the tendency to radiate from the center toward the circumference. In the lustre of gold something is added that, despite all radiation, remains fully master of itself. There is an "inner heaviness" in the yellow of gold; it gives itself weight. In colloidal solutions gold shows an enormous colorfulness; it can be greenish-blue, violet, reddish, violet-blue, or pure rose. A dilution of 1:100 million still makes water visibly purple. The beautiful red glass in cathedral windows is due to the addition of a trace of gold when the glass was poured.

In material form gold can be stretched to an incredible degree without losing its cohesion. When beaten into the finest leaf, it reaches the point of translucency, allowing the light to shimmer through bluish-green as through the leaf of a plant, but still retaining its pure yellow glow when the light falls directly upon it. Gold is the most ductile and malleable substance known. We can make it into a film less than a ten-thousandth of a millimeter thick. A wire almost thirty-five miles long can be made from a single ounce of gold. This reveals the strongest forces of cohesion and the highest capacity of extension. More than any other substance gold can change from a three-dimensional body to a flat, two-dimensional condition without crumbling. This plasticity, this "inner fluidity," speaks of special forces standing behind its physical nature. In addition, we must consider its excellent conductivity for heat and electricity. Nevertheless, while some other metals, such as lead, tin, or quicksilver, lose all resistance to electric current under extremely rapid cooling with liquid helium and become "super-conductive," gold does not let itself be altered thus by cold.

The high density of gold, which is nineteen times heavier than water, as well as its high atomic weight, again shows the intense power of its material life.

Thus gold is a metal between light and gravity, like no other. It is both sun-bright and earth-heavy. It bears within itself three principles: light, elastic fluidity, and weight. The ancients sought three principles in all earthly substance in order to show, by the way they worked together, something of the nature of matter: a principle of the cosmic light nature (sulphur); an earth-like principle (salt); and, connecting these two, a rhythmical, half-cosmic, half-earthly one (mercury). In their alchemistic procedures they sought to separate these three principles, and found that in gold this separation was the most difficult to achieve, gold being even harder to destroy than to build up. This expresses in a striking way the inner cohesive force of gold.

SALT, MERCURY, SULPHUR.

From what has been said, the reader will realise that these three names do not simply mean salt, quicksilver, and sulphur in the present sense, but three force-principles that were felt to be present in every earthly substance and process. First, let us remember that the solid, the liquid, the airy, and the warm were regarded as the four earthly conditions. The processes that led the liquid into solidity were seen as salt. Mercury was felt to be active in every rhythmic combination, in the mutual interpenetration and separation of the aeriform and the liquid. Sulphur lived wherever the air was penetrated by heat and evaporated, that is, where combustibles arose and broke into fire. Within the mineral kingdom these three substances were merely special representatives of the three principles. But they were found also in the plant kingdom, in the root processes, the leaf activity, and the blossom formation. In man, we may consider the head and nerve processes to be salt, the activity of the middle rhythmic man to be mercurial, and the metabolic processes to be sulphuric. As modern men we can still see something sensible in such views, which were meant not simply in a static-material but in a dynamic sense. The more our perception of nature rises from the observation of the static, of what has already become, to a living grasp of the dynamic process of becoming, the greater will be our understanding for these ancient concepts. This is true even where we are proceeding on new paths.*

The Chemistry of Gold.

Gold is so noble that it is hard to dissolve or combine it. We must use aqua regia (a mixture of nitric and hydrochloric acids) to change it into a salt. Then it forms gold trichloride, $AuCl_3$. Among the metal chlorides, notably among our seven metals, this resembles most closely iron chloride, $FeCl_3$. As a salt, gold thus inclines toward iron. For metallic gold, however, the best comparison is copper, the base-born half-sister. Gold thus stands, even in the inorganic, in a certain sense between iron and copper. This can be seen clearly in the diagram of our seven-pointed star. Our observations on the physiology of the metal processes in man will also show gold to be between iron and copper.

Chemically, gold is both uni-valent and tri-valent. But the uni-valent condition is the less stable. The "valency" of the metals is a property that also fits into the oft-mentioned heptagonal diagram. Iron is essentially bi-valent, and tri-valent; copper, uni- and bi-valent; silver, uni-valent; tin, tri- and quatro-valent; gold, uni- and tri-valent; quicksilver, uni- and bi-valent; lead, bi- and quatro-valent.

"Valency" expresses whether a metal is already saturated (neutralized)

---

* Such views are in harmony with the latest endeavors in biology and medicine to overcome "static" thinking and to break through to a grasp of the dynamic, the functional. A "functional pathology" is already a clearly discernible goal.

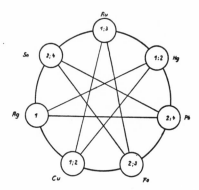

by a single combination with a uni-valent element, such as hydrogen, or is capable of further combinations. Multiple valency shows greater inner mobility and complexity.

Another solvent for subtly distributed gold is—in the presence of oxygen—a solution of potassium cyanide. This has enabled us to win from the rock some of the finely disseminated gold by which the earth is delicately tinged. Prior to the middle of last century man could extract gold only where nature itself had concentrated it. Since 1887, however, when the process of cyanic leaching was discovered, ores can be worked whose gold content may be as little as the sixth decimal potency, or one part in a million.

GOLD MIRRORS.

A form in which the "light aspect" of gold is especially manifest is the mirror, where the gold backing, almost wholly two-dimensional, most nearly approaches the light. It is all lustre, transparent, thin as a breath, and at the same time all color. To make the mirror, the metal is first vaporized. As a shining blue-green "steam," it is a "being of color." Out of this coloring it condenses to a golden film on the cold walls of the vessel in which it is vaporized. In this form it is very pure. It can be brought into this condition only by the hand of man.

Contrasting with this is natural gold, the qualities of which are entirely bound up with the deposits where it is found. Abyssinian gold, for example, is extremely pure. The gold of the Urals is distinguished by a certain admixture of platinum. In antiquity Asia Minor yielded a white gold, rich in silver, which was known as electrum.

GOLD IN HISTORY.

Heretofore we have considered the natural side of gold. But man wrests the metals from nature and brings them into the realm of civilization. There, to paraphrase Goethe's statement about the colors, they must go through their "deeds and sufferings" according to their character. Each metal experiences, and provokes, a different destiny.

In ancient times gold served only cultic purposes; it belonged to the priest-king, who kept it in trust for the sun-god. In antiquity Egypt was the land richest in gold (and poorest in silver). It derived its gold from the desert region on the Red Sea and from Nubia. (Nub in the Egyptian language means gold.) Later the gold stocks of the world flowed to Persia. Cambyses captured the gold of Egypt, Darius that of India. The first minted gold coins are said to have been Persian, with pictures of the gods stamped on them. After the fall of Persia all the gold of the ancient world streamed into the hands of one man, Alexander the Great. He was one of the first who stood entirely upon the strength of his own personality. He was also the first to adorn gold coins with a human image, his own. He inaugurated a redistribution of gold throughout the then known world. After his death it flowed into all parts.

Rome got most of its gold from Spain, the America of the old world, keeping 60,000 slaves continuously in bondage at the mines. Plunder also did much to fill the treasury. In Caesar's time, for example, the immense temple treasure of the Gauls, accumulated for centuries in honor of the gods, was carried off to Rome. Its value is estimated today at more than a billion gold dollars. But Rome could not retain this gold. It drained away uninterruptedly to the East, to pay the border legions, to finance the constant wars, to buy goods from the Orient. The decline of Roman power was partially caused by the inability to replenish this constant outflow.

With the rise of Christianity, slavery came to an end. Mining also stopped and the mines remained closed for centuries. The early Middle

Ages were poor in gold. Then the northern gold deposits, the Alps, the Carpathians, the Bohemian mountains, gradually came to the fore. Meanwhile a new class of free miners came into existence. These miners pioneered in the German movement to the East, founding the first German cities in the Sudeten, Silesia, and Hungary. Gold was a blessing in many ways, but soon it called forth a dark destiny. The Templars, in the Middle Ages, became the lords of gold and the bankers of Europe. Philip the Fair, gripped by an insane thirst for gold, seized the Templars' treasures and cruelly destroyed the entire Order.

When America was discovered, or rather rediscovered, immeasurable stores of gold, accumulated throughout centuries in honor of the gods and formed into cultic objects and images of the Sun, were plundered from the Aztecs and the Incas with unspeakable crimes. The lust for gold played a large part in the discovery of the New World and is not altogether without influence today.

Then the search for Eldorado shifted to the desert continent of Australia. The gold discovered there enabled it to complete its rapid rise from a penal colony to a modern state. But it has not yet made this continent culturally creative. Whoever sees merely material value in gold suffers a peculiar fate.

In 1816 England introduced the monetary gold standard. The gold impulse now seized the economic life and this, contrary to its original purpose of selflessly serving human needs, became increasingly the instrument of the most powerful state ever known in the world. African gold combined with this world power. Now the process of cyanide leaching began to play its role, attacking the "homeopathic" gold stratum of the earth.

Gold now flowed toward the West, just as during antiquity it flowed toward the East. With this change in direction came an equal change in use. Gold no longer serves religious or cultic purposes. The largest pile of gold in history lies in the vaults of a North American fortress, where the naked ingots help to propel the wheels of a mechanical economic order. What will happen to these tremendous quantities of gold?

Among the deeds and sufferings, blessings and curses of the precious metal, it once formed the implements and images of the gods; later, it was a noble material for the artist; now it is an anonymous factor in the abstract world of economic thinking. But are these two realms not also con-

100

nected with the sun? The one, forever life-engendering, making the plant forms into images of constantly renewed life; and the other, in which all life ceases, having its image in the desert? The creative and the demonic—both pertain to the sun as well as to gold. And they pertain to gold as to no other metal.

GOLD AND THE HUMAN ORGANIZATION.

One might think that a substance so rare and noble, which combines with no other but remains unassailably itself, would have nothing to do with man, at least so far as the latter is physical. On the contrary, however, gold has strong and unequivocal effects on man, in health as well as in sickness. But what is essential here is not gold *matter* but gold *dynamics*. When a healthy person is given minimal doses of fine gold triturations (powder) over a prolonged period, as is done for the so-called homeopathic test-pictures, we find characteristic effects due only to gold. We will discuss these effects in the remainder of this chapter.

The blood circulation is primarily affected. The blood floods into all the organs; into the head, which may become red, swollen, hot; into the chest, causing lung congestion, shortness of breath, anxiety, fear, cardiac oppression, strong heart palpitation and faltering of the heart activity. The blood pressure rises. The circulation may also become congested in the legs, in all the lower organs, resulting in edema. The sense organs, too, overfill with blood. Photophobia (aversion to light) and hypersensitivity of hearing appear. The skin is attacked by a rash, eczemas and skin inflammations may occur. But the heart above all, as the central organ of the blood circulation, shows the influence of the gold. This is why gold can be a powerful and important heart remedy.

In numerous lectures Rudolf Steiner revealed much of the nature of the metals as they appear to a way of cognition that can ascend from physical to spiritual observation. About gold he said, "From external analogies it is assumed that the ancients saw in gold a representative of the sun. But it was not merely a superficial play of analogies that considered the sun something precious in the sky and gold something precious on earth. Nothing is really too stupid for modern man when it comes to imputing stupidity to the ancients. When they looked at gold, with its self-possessed radiantly yellow color and its modesty and dignity, they actually felt something related to man's entire blood circulation. Confronted with the quality of gold, they felt: you are within this, here you can feel yourself. And by virtue of this feeling they understood the nature of what is sun-like. They felt the kinship of the quality of gold with what came from the sun and worked in the blood of man." *

Connected with the effect of gold on the circulation is an effect on the warmth organization. It is useful for chills, heat flashes, night sweats. Gold salves, which Rudolf Steiner suggested for the treatment of lupus, have an effect similar to that of irradiation with sun rays. But they also act harmonizingly on the heart activity.

The nervous system likewise experiences the violent impact of the blood forces under the influence of gold. These forces may cause inflammations by pressing into the area of nerve activity, which is normally in such contrast to the activity of the blood. Various neuralgias are the result.

The digestive system reacts with salivation, tooth-ache, inflammation of the gums, in the upper region; with pain in the middle region; and with meteorism (abdominal distension) and constipation in the lower. Our bones and muscles, all that gives us support and movement in life, hurt. We feel as though "beaten up."

The blood unconsciously "hungers" after what the upper sense-nerve and breathing region, and the lower metabolic region, can convey to it. In its motion it seeks for light and air from above and nourishment from below, and in this motion it finds its satisfaction. Gold intervenes in all these relationships. In our observations of the inorganic world, we noted the position of gold between light and gravity; the blood, with the heart as its center, lives in the body in a similar polaric tension between light and air, and weight and the forces of matter. Gold combines neither with silica nor with oxygen, but it penetrates quartz, which in man stimulates the sense activity, and it accompanies iron, which in man lies at the foundation of breathing. Gold is not receptive to sulphur, but it tinges the sulphuric metals, and sulphur works in the metabolic forces of the lower organization. Under the influence of gold the blood forces swing strongly back and forth between the sense-nerve and breathing organization and the limb-

---

* R. Steiner, *Initiate Consciousness.*

metabolic organization. The former, as we have seen, combines with the effects of lead, tin, and iron; the latter, as remains to be shown, with those of copper, quicksilver, and silver. Not only in nature but also in man, gold stands harmonizingly in the center between the two polaric groups of metals or, to be more accurate, of processes pertaining to them.

But before all the above physical symptoms manifest themselves, there occur distinct changes in the state of the soul. Melancholy, fear, even despair, may grip the soul, going as far as suicidal tendencies. Feelings of inferiority, self-reproach, and religious obsession, have been noted under gold treatment. It can become a "poison for body and soul" (Stauffer). The feelings rather than the reason seem to be affected and disturbed. (Within the threefold organization, according to Rudolf Steiner, the rhythmical system is the carrier of our feelings; the nervous system and the brain are the carrier of our thoughts.) One becomes irresolute, quarrelsome, unable to bear contradiction. Thus, too much gold brings man's spirit into disagreement with his body. In a certain sense it is the mood of Faust in the scene in his study when, because of torments of conscience, despair, anguish, and frustration, he seeks escape into spiritual regions through suicide. Luciferic spirituality, which shuns all earthly tasks, is what Faust is seeking at this stage of his life.

If we are serious about overcoming materialism, and if we look for the creative spirituality behind all material objects, we must expect to find a definite differentiated spirituality back of every earthly substance and process. Having this clearly in mind, we will admit the validity of spiritual research as we admit the validity of scientific research. In our own studies we will elaborate the concrete accounts of the spiritual researcher: that gold, for example, is a basic mineral substance that received its essence from the sun itself; that Luciferic Spirits of Wisdom sent to the earth etheric streams which, in the course of evolution, have become gold as it is known today.* Etheric life-streams from the universe have secreted matter out of themselves just as the life-streams in the body, for example, secrete hair or nails into the lifeless.

An excess of gold can call forth the above symptoms in a healthy person. If gold is properly prepared, the forces manifest in these symptoms are transformed into such as can heal a sick person in the regions where his own "gold process" is disturbed.

---

* R. Steiner, *The Spiritual Beings in the Heavenly Bodies and in the Kingdoms of Nature.*

VII

COPPER

It needs only a quick glance to see that copper and its numerous ores are extraordinarily colorful. The pure metal can be rose-red, sunrise-red, reddish-yellow, or brown-red. In the transparency of extremely thin sheets it is a shadowy blue-green, complementary to the orange-red of the polished metal. Blue, green-blue, green, and violet tones appear in chemical solutions of its salts, and a world of color reveals itself in the copper ores, a joy to the eye such as no other metal provides. They are the first to attract our attention in entering a well ordered mineralogical exhibit. Copper pyrite shines golden yellow, malachite and olivenite are green, azurite is blue, covelline is violet-blue, and the spotted copper pyrite (bornite) shimmers in all colors of the rainbow.

The impression made by a lustrous piece of copper is pleasant, warm, and friendly. A copper mirror reflects our image in the colors of radiant health. It is a pleasure to handle a copper tool or utensil. We can readily grant copper the rank of a humbler half-sister of the noble gold.

In keeping with its semi-precious nature, copper is often found as a pure metal; sometimes in dainty fern-like forms, at other times in great lumps and blocks. But most copper appears in the form of various ores.

The forms of its important deposits and the material composition of its ores show us the connections between copper and the evolving of the earth. A chapter of "earth physiology" arises before us, the counter-image of which we shall find in man as we study how his organism makes use of copper. We can go even further if we see how copper behaves toward various other earthly substances and processes.

104

From the earliest ages copper has been immersed in a world of sulphur processes; it has a great affinity for sulphur, and the most important copper ores are sulphur combinations. They prefer the deep-seated, quartz-poor, basic igneous rocks, the melaphyres, etc., avoiding the quartz-rich, granitoid rocks (where tin is more at home). This siliceous variety of the granites has in the course of time thrust itself strongly upwards to the surface, as an inorganic counter-image of the fact that in the organic world silicic acid pushes outward in skin, hair, and sense-organs. In this silicic region copper is not at home.

We do not find it in the "acid" upper regions where the oxygen-bearing tin ores predominate, but in the "alkaline" depths among the sulphides. The commonest and most important copper ore is chalcopyrite, a copper-iron-sulphide. It is significant that nature permits the encounter of two polarities, copper and iron, in the sphere of sulphur. It mentions copper and iron in the same breath, as it were, a fact of earth-physiology whose true significance dawns on us only when we know the roles that copper and iron play in human physiology. Besides this chief ore, the alkaline igneous rocks (gabbros, etc.) shelter, as "copper formations," many related sulphur ores, thus varying in many ways the iron-copper combinations. Among these are bornite, chalcocite (copper glance), pyrite, and magnetic pyrite (pyrrhotite). In tetrahedrite, silver and quicksilver join in; lead is found in bournonite; and in tin pyrite there is, of course, tin. In addition, the "fahl ores" contain two sulphur-friendly elements, antimony and arsenic (tetrahedrite and tennantite).

Thus copper and sulphur were locked together in the earth in primeval times. Wherever these ores come to the surface now, the present atmospheric conditions transform them in a highly characteristic manner.

COPPER AND OXYGEN, WATER, AND CARBON DIOXIDE.

The weathering of such parts of a copper ore deposit as appear on the surface is, to begin with, an oxidation. The air and water cause the iron to rust and the sulphur to "burn" and acidify. Water-soluble metal salts are formed, which seep down, take hold of the deeper-lying and as yet un-touched metallic sulphides, change them, and exhaust themselves in the process. This creates a boundary to their activity, a kind of "diaphragm" separating the lower unchanged sulphur zone from the upper oxygen-per-meated, water-saturated oxidation zone. We have before us an air-per-vaded "above," a sulphur-pervaded "below," and a middle region in be-tween. In this middle region the precious metals in the deposit accumulate. It is called the cementation zone. In it the concentrations of the pure copper sulphides that have become free of iron, copper glance and copper indigo, also occur. Toward the surface, in the region where oxygen predominates, we find the oxidized copper minerals such as red copper ore (cuprite). Because the iron also oxidizes and rusts here, this outcropping is called "iron hat" or gossan. Right on the surface, however, under the influence of carbonic acid and moisture, there occurs the final and most sta-ble form of all these metamorphoses, the copper compound that is most appropriate to the earth's present condition. This is the basic carbonate of copper, azurite and malachite, compounds in which water is present as well as the airy element of carbon dioxide. Under present conditions all copper turns to malachite in the end. Ancient periods of the earth gave copper the form of copper-pyrite; the present period aims at malachite. This min-eral, which we find in knobby, kidney-shaped, fibrous masses of a beautiful emerald green, is the last phase of the copper process in the mineral earth. The "metabolic processes" of the earth drive copper out of the realm of sulphur into that of oxygen and carbon dioxide.

The Realm of Copper.

If we study the distribution of the known copper deposits, to see how our earth is "copper-penetrated," we are immediately struck by a prepon-derance in the western continents, a balance in the central, and a meager-ness in the eastern continents. North America contains about 36%, South America 25%, Europe 11%, Africa 18%, Asia 9%, and Australia about 1% of the world's wealth in copper. Most copper lies around the Pacific Ocean, which also has many volcanoes on its periphery. 77% of the American copper lies in the young folds of the Western Cordilleras (Andes, Rockies, etc.), which still exhibit fairly strong volcanic activity. If we could relocate the earth's axis so that the north pole fell in the area of the Belgian Congo and Northern Rhodesia and the south pole in about the middle of the Pacific Ocean, the polar cap would embrace the great African copper fields of the Katanga area, etc., and the equatorial zone would be a copper belt running through both Americas and embracing the important Asian and Australian deposits. It is characteristic, however, that the present poles are not "copper oriented" but, as indicated in Chapter 5, iron-oriented. The "copper pole" is situated in the African tropics, the "copper equator" around the Pacific Ocean.*

So much for the occurrence of the crude tangible quantities. In subtle distribution, however, copper may be found everywhere. The earth layers accessible to us are permeated by copper to an extent comparable to about the 4th decimal potency, i.e., one part in ten thousand.

THE PURE METAL.

The relation of copper to the world of light is apparent in the previously described colorfulness, in which it is unsurpassed by any other metal. The active colors, such as yellow, orange-gold, and red, arise in the metal itself, in cuprite, copper glance, and copper pyrite, as well as in the oxygen and sulphur-iron combinations. The passive colors appear in the copper salts and carbonate ores. (Very intense blue pigments, as well as indigo, show a coppery sheen on their surface.) Heat finds in copper its best metallic conductor, next to silver. It offers the easiest passage to electric currents. It shares with the precious metals an inner plasticity and fluid ductility. But it is much harder than these, tough and solid, and in this is closer to iron. If we compare the various metals in these respects, the following distinctive grouping results:

Metal	Silver	Copper	Gold	Iron	Tin	Lead**
Electric conductivity	100	93	67	16	14	7.8
Heat	100	91	74	13	15	8
Tensile Strength	18	22	10	25	2	9

* Compare G. Wachsmuth, *The Evolution of the Earth*, especially Chapter III.
** Quicksilver, as a *liquid* metal, does not belong in this series.

This table shows a polarity among the metals in their inorganic aspects. The malleable heat-conducting metals fall on one side and the less solid, poorly-conducting ones on the other. Gold stands in the center, mediating between copper and iron. In the metal processes in the organic realm, these polarities will be found in a much more perfect form.

Copper has a happy relationship to the element of sound. It is a "euphonious" metal. Many wind and percussion instruments bear rousing witness to this fact.

Copper melts at 1083° C. and vaporizes at 2510° C. Its steam shines with the same beautiful green-blue light that appears when we look through the finest leaf. In melting, it reveals its affinity to the airy realm by the remarkable phenomenon of greedily sucking in gases, such as hydrogen, carbon monoxide, and sulphur dioxide, only to expel them again in spattering little explosions as it hardens. Silver does the same with oxygen.

The Chemistry of Copper.

Because of its versatility in transforming and combining, the alchemists called copper "meretrix metallorum," the harlot of the metals. It displays a certain reserve toward oxygen and water, resisting the activities that these substances develop in the life of the earth. On the other hand, the combined effects of dampness and carbon dioxide in the air overlay the red metal with the noble green rust known as patina, and in this colored sheath (of a malachite character) it can defy time. It combines easily with the other elements, so that even weak acids such as vinegar dissolve it, but it returns just as easily to its metallic condition, being "semi-precious" in this respect also. Like iron, it fluctuates between two conditions of valence, the uni-valent silver-related compounds, the cuprous salts, and the bi-valent cupric salts. Again like iron, in certain lower animals copper can be a breathing metal.

In the deep blue solution of copper sulphate a phenomenon occurs that can also be observed in alum solutions. The light that traverses it is robbed of its warmth and comes out "cold." Copper chloride is able, with the addition of a little blood or plant extract, to form its fine needle-like crystals into such characteristic groupings that the shapes thus made visible may actually be used as indicators of the formative forces. We refer here to the significant work of Ehrenfried Pfeiffer and his pupils.* No other metal has shown such sensitivity in this realm.

COPPER IN THE PLANT WORLD.

When carefully examined, our metal proves to be widely prevalent in the plant kingdom; in fact, it is present everywhere. It appears mostly as a "trace element," thus showing that what is essential is not its substance but the dynamic impulse that it arouses. It is present in our important food stuffs, for example, four parts in a million in bread, two in potatoes. Copper deficiency produces serious plant diseases in grains and legumes, which may be cured by proper doses of copper. Chlorophyll defects, leaf necrosis, weakening of the generative as compared to the vegetative development, all these are symptoms of copper deficiency. The efficacy of Vitamin B seems to be connected with the presence of copper traces, and the amount of this substance in the tissues of plants and animals is said to run parallel to the amount of copper. Some plants, especially in the carnation family, are stimulated by copper and thus serve the miner as "copper indicators." Conversely, lower plant forms are suppressed by the effects of copper. Unicellular plants, low types of mushrooms, algae, for example, will be killed by the oligodynamic effect of a copper coin placed in an aquarium in which they are present. More highly organized life will seldom suffer from this.

COPPER IN THE ANIMAL KINGDOM.

Our observations of the plants showed that copper was important in the organic realm. More light is thrown on the copper process when we ascend to the animal kingdom. It was astounding when marine molluscs, mussels, and snails, (including fresh-water and even land snails), as well as cuttlefish and crabs, were found to breathe with the aid of a blood pigment

---

* There is a wealth of material. We will cite only E. Pfeiffer, *Sensitive Crystallization Processes* (1936); H. Kruger, *Copper Chloride Crystallizations: A Reagent towards Formative Forces in Organic Life* (1950; A. Selawry, *New Results in the Field of Copper Chloride Blood Crystallizations* (1949).

that contains not iron but copper. The hemoglobin of warm-blooded creatures is replaced by the copper-bearing hemocyanine. The characteristic feature of the above animals is that they maintain their inner body in the completely soft plasticity of live proteinaceous substance, and expel everything hardening and form-giving into an outer skeleton or shell, as the oyster, for example, expels all of its calcium. They belong to a fluid world and must extract their oxygen from the water; they are water-breathers and not air-breathers. To this they owe their dull consciousness, their voicelessness, the lack of inherent warmth. Albumen-like plasticity and metabolic organization predominate in these creatures; they enclose themselves in their rigid sheaths. Nature, in answer to our quest for the mystery of copper in the realm of the living, places the picture before us of the form of the mussel and the snail.

But there is also a counter-image in the form of the bird. For there are birds, such as the touraco, that take copper into themselves but, instead of retaining it in their blood, secrete it in their feathers as a red coloring, turacin. This pigment, containing 6% copper, is here incorporated into the flying rather than the breathing process. With their beautiful blue, green, violet, and red colors, these magnificent birds present a living image of the essence of copper.

A complementary polarity is the snail, with its fluid albumen-like body protected from every hardening; with an eye that can be extended from the body like a limb; gliding over the solid ground in its own slime; artfully moving everything solid and calcareous towards the outside. It breathes with copper. The touraco, on the contrary, with its deeply imbedded air-organization, desiccated at the periphery, with a finely wrought inner skeleton, breathes with iron. Its body is too warm when compared to the cold snail body. Along with the air, the bird has been pervaded by a quickening soul nature, which uses this air as a voice to express its soul. What in the birds makes such a rich outward show as movement, color, and voice, is in the molluscs merely metabolism, closed in upon itself. Here the copper process is the breathing instrument of the animal's astral body. In the bird it yields its activity to iron and devotes itself to color in the air element, in the organ built for the air.

It would be interesting to examine the copper content in feathers of other birds, for it would be strange if this metal were not present also in them, though perhaps in very fine doses.*

In our discussion of iron we said that there is a strange and mysterious link connecting the synthesizing pigment of the plant and the breathing pigment of the molluscs, crabs, and higher animals. Added to this is the feather pigmentation of certain birds. We have here similar basic organic substances, the porphyrines. These turn into chlorophyll when absorbing magnesium; into hemocyanin when incorporating copper; into hematine when they lay hold of iron. Iron and copper are not contained in the chlorophyll, but they must be present in its "environment," for without traces of iron no chlorophyll can come to birth, and as we have seen, copper must also be present. As soon as copper or iron is taken into the porphyrines, we advance from chlorophyll to blood. Into the merely vegetative life there enter, at the same time, higher members. The plant ascends to the animal. For the management of copper or iron *inside* an organism, higher members are required than those of the plant. The plant does actually possess members higher than its physical and formative forces (etheric) body, but they work *outside* the plant, never manifesting from within and remaining entirely supersensible.

This indicates that copper is the instrument of a principle that goes beyond the plant nature and pertains only to the animal and to man. Therefore we may expect a further revelation of the nature of copper as we progress from the plant to higher forms of life.

COPPER AND COPPER PROCESSES IN MAN.

The entire human body contains about 0.2 to 0.3 grams of copper in organic combination, as against about 4 grams of iron. This copper is not an accidental affair; it is a normal and indispensable ingredient, as we shall see. It is unevenly distributed through the organism, with the various tissues containing the following number of parts per million:

brain	pancreas	heart	muscles	bones	kidneys	lungs	spleen	liver
3	3	3	3	3	2.98	2.5	1.78	7.5

---

* In the Tabulae Biologicae X,210 (1935) there are articles giving a copper content 0.76 to 1 part per million in bird feathers. There is an average of 0.4 parts per million in the human body.

111

The liver is thus richest in copper, especially the fetal liver. In the brain the richest part is the *substantia nigra,* which has melanin-containing nerve cells. (The formation of melanin appears to be promoted by a protein-copper combination.) Here we meet the copper process again as it presses to the periphery, in connection with pigment formations. Dark-haired animals (rabbits), have more copper in their hair than light-haired ones, and in spotted dogs and cats the pigmented hair contains more copper. It is also remarkable that tumorous tissue contains more copper than healthy tissue.

Copper has especially interesting relationships to the blood. It is necessary for the formation of blood. Certain forms of anemia can be cured only if copper is used to help with the incorporation of iron into the hemoglobin. Copper itself appears mainly in the blood serum, in which a certain copper level is always carefully maintained. Here it works together with the serum iron in a way that varies with the sexes. The serum iron level in the male amounts to 118 gamma %, while in the female it is only 88 gamma %. The serum copper level on the other hand amounts to 106 gamma % in the male and, at 107 gamma %, is practically the same in the female. The balance between copper and iron thus tends, in the woman, in the direction of copper. The male is, in this connection, richer in iron. Copper is only loosely bound up with the proteins, but the level of the serum copper is tenaciously maintained in hunger, fever, poisoning, etc. During pregnancy it rises to 280 gamma %. In infections, the interplay between copper and iron is shifted, the iron level sinking and the copper level rising. When healing takes place, the levels approach each other again. again.

The more active the metabolism becomes, the higher the copper level seems to rise. Hypofunction of the thyroid gland lowers not only the basal metabolism but also the copper level. Excision of the thyroid gland has the same result. In hyperthyroid function, on the other hand, such as Basedow's disease (exophthalmic goitre), there is a marked increase in the copper level.

It is remarkable that the level of the serum copper is abnormally high in certain mental diseases, such as schizophrenia or manic depressive insantiy, and also in epilepsy. This should induce us to follow up Rudolf Steiner's suggestion that the so-called mental diseases are in reality physical illnesses and that some kind of organic degeneration is always lurking in the background.

Abnormal increases in the copper level also occur in diseases of the liver and gall bladder. In jaundice particularly, the bilirubin level and the copper level of the serum run parallel, rising to abnormal heights as the illness develops and returning to normal when the illness is cured.

Thus man has a firm relationship to copper by way of his inner body, but many characteristic phenomena are produced by copper from without. To begin with, the handling of this metal causes no occupational diseases, in contrast to the often serious chronic illnesses that result from minute quantities of such metals as lead, chromium, quicksilver, uranium, etc. On the contrary, copper has many favorable effects. Copper workers are said to have shown special resistance in cholera epidemics, seldom succumbing to this disease.

But if small doses of a copper salt are taken by a healthy person regularly and over a prolonged period, he will lose his "copper equilibrium." There occurs an over-emphasis of the copper process that in the normal course of events must work in harmony with many other processes, notably that of iron. These effects appear in the so-called test pictures of the homeopaths. Hugo Schultz describes them as follows:

> Interestingly enough, psychic symptoms arise, such as melancholic depression, fright, fear of death, weakness of memory, inability to retain thoughts. (We may recall the disturbed copper level in genuine mental illnesses). Added to these psychic disturbances are others, such as dizziness, heaviness, rush of blood to the head, that indicate how the consciousness-bearing factors, above all the ego, are pressed out of certain bodily regions. The principle guiding the blood distribution is no longer entirely in order. The pulse becomes accelerated, hard, short; chills run through the entire body, the warmth-organization is upset. The lungs become too full of blood. Pressure, tension, shortness of breath are experienced. Not only the blood but the entire system of fluids flows, insufficiently warmed, toward the upper organization; outbreaks of perspiration, congested mucosa, laryngeal catarrh, and hoarseness bespeak this fact; as does convulsive coughing, especially at night when our higher members, particularly the astral body, are loosened from our body. But the effects of copper also

extend into the digestive system; the pharynx becomes constricted, swallowing is difficult, nausea and gastric catarrh appear, even vomiting. There is constipation, followed by diarrhea. The liver area becomes painful, there is urinary frequency with small voidings. The inability to control the metabolic processes shows up as eczema. Muscular spasms and painful contractions indicate the effects on the limb system. (*Wirkung und Anwendung der unorg. Arzneistoffe.*)

Copper and its combinations are among the oldest remedies. The Ebers papyrus (1500 B.C.) mentions their efficacy in cases of chlorosis egyptica (a form of anemia known as "greensickness") and in swellings of the glands. The writings of Hippocrates mention pneumonia, empyema, and hemoptysis, as sicknesses that can be helped by copper. Paracelsus attempted to cure mental diseases, epilepsy, hysteria, as well as pulmonary diseases and syphilis, with copper. Rademacher termed copper a universal remedy, which he used in pneumonia, tuberculosis, rheumatism, dropsy, and dysentery. It is beneficial in diseases of the intestines and against intestinal parasites; expanding from these areas, it was found to be effective also in the neighboring regions of the inner digestion, the area of the portal vein and the liver. Copper is useful when the pathologically displaced metabolic activity, in a metamorphosed form, works improperly into the "foreign" realm of the lung processes as in icterus (jaundice), chronic pneumonia, pleurisy. Chronically weeping eczemas occur when these pathological metabolic activities break into the sense-nerve domain. The efficacy of copper against certain forms of rheumatism represents the same process in a different direction. Cholera and certain forms of epilepsy form the upper boundary to the use of copper in the entire human organism.

Copper Therapy as Expanded by Spiritual Science.

We will find the leitmotif of the copper influences if we understand their ways in nature, and then imagine these ways extended into the realm of man.

Let us summarize into a unified whole what so far has been described in many details and perhaps somewhat extensively.

In its ores copper appears as something primarily sulphur-connected. It

114

inclines toward sulphur. It is thus able to fit itself, even in the organic world, into processes that may be properly called sulphuric, since they represent an association with the sulphur substance. It can "easily join the course that sulphur takes within the organism." * As a consequence it can be "easily integrated into the activity of protein within the human body," and combines easily with these upbuilding proteinaceous processes. It "supports the ego organization that becomes weakened in the digestive tract." It helps the higher principles, the ego and the astral body, to take part in the digestive organization. A hypersensitivity of the astral body in the realm of digestion (shown by painful cramps) would indicate that the etheric body was not adequately performing its own appropriate functions in this part of the organism, but was imitating the functions of the astral. Copper divests the astral body of this hypersensitivity. It helps the ego organization generate warmth in the digestive tract whenever this warmth has become deficient. The beneficial effects of a massage with copper ointment belong in this realm.

In outer nature copper is not only confronted by sulphur, but is closely associated with iron. In the organism this comes to expression in copper's basic tendency to guide the upbuilding protein metabolism in such a way that this enters the circulation and strives to penetrate the breathing. Copper "supports the astral forces in taking hold of the circulatory system." Here it gravitates away from sulphur and toward oxygen, and either turns itself into respiratory pigment in the lower animals (in whom the astral body has only lightly entered) or, in the higher animals and man, delivers the metabolism into the hands of iron. It promotes the formation of red corpuscles. With this, it yields to iron the higher region where the organs of breathing and consciousness are active. Warm blood is formed; breathing of the air begins, making way for the voice. Calcium is taken into the body and shaped into an inner skeleton. We now have before us, in man, a being who not only has a sentient soul, but an "intellectual soul" and a "consciousness soul." Copper remains strong in the "lower region" of the circulation, in the veins. It has given the arterial region up to iron. (We have shown how in present-day nature copper tends toward carbonic acid and hydration as malachite). The organization rises out of the moist and

---

* Quotations are from R. Steiner and I. Wegman, *Fundamentals of Therapy*; and R. Steiner, *Man as a Symphony of the Creative Word.*

cool region of the unformed proteinaceous nature, which was not able to absorb and shape the mineral nature but could only reject it, by lifting itself from the copper process to the iron process. The bird even carries the "copper blood" of the molluscs in its feathers; he has pushed out the excess of the copper process that belongs to him as a creature with a high basal metabolism; he carries the copper into the regions of air, light, and color. The bird carries physically in his feathers what man carries etherically in his thinking. Thus man can manifest copper in a special sphere. Much contact with copper makes him "sharp" in the physical world; for example, he can acquire the ability to find the book he wants without interminable searching in the library, and to open it immediately at the desired place. This was remarked by Rudolf Steiner.

From the above we will understand why all manner of venous congestions, varicose veins, etc., are an important field of copper therapy.

COPPER IN THE CULTURAL SPHERE.

From outer nature copper has been admitted into the inner man, or rather retained within him,* in order to serve the unfolding of man's higher members and to enable the ego (in cooperation with the astral) to work into certain bodily processes. It has also been received into, or restored to, man's cultural activities, and it has thereby acquired a special destiny.

Man took hold of copper during the third post-Atlantean cultural epoch, whose task it was to develop the sentient soul. We find the first copper implements in Egypt, about 3900–3600 B.C. The period of the early dynasties already shows extensive use of them. From about 3100 B.C. on, all important utensils were made of copper, and the people lived in constant contact with it. This is followed by an ancient Bronze Age, approximately from the beginning of the pyramid period, 2700–2300 B.C. Then comes another Copper Age, which is followed by the main Bronze Age, at about 2100 B.C.

---

* We remind the reader of the description in the opening chapters, according to which the nature kingdoms below man are excretions, sloughings-off, from the process of human evolution.

The leading cultural peoples were largely adherents of the Venus-Astarte cult. The Phoenicians became masters of the art of casting. The weapons of those ancient peoples, the helmets and shield-bosses of the ancient Sumerians, for example, were made of copper. Egyptian miners worked their stone with hardened copper tools, as did the masons of the pyramids in squaring those mighty granite blocks. The Romans introduced an Iron Age, but they still extracted enormous quantities of copper from the well-nigh inexhaustible Spanish mines. The great migrations brought mining to a standstill. In the Middle Ages, mining arose in Germany, and the Mansfeld copper-slate mines supplied Europe with the red metal for centuries. We may always recall the warm-tempered Middle Ages as a time in which, though iron assumed an ever-growing importance, the people were nevertheless in daily contact with copper, in every doorknob, in cooking pots, candlesticks, lamps, etc. To this day the Italians cook their polenta in copper kettles. Certain countries, such as Japan, still lived in a Copper Age until the middle of the last century.

The ancients did not ascribe the metals to the forces of the earth. They experienced the metals as alien to the earth-forces, which constantly seek to annihilate them and therefore could not possibly have created them. They saw the metals as gifts from heaven (which today would be admitted only with regard to meteoric iron) and as the results of influences that in primeval times worked on the earth from the cosmos and the planets, then became concentrated into matter on earth. Copper was connected with forces of a super-material nature radiating from the planet Venus (the stars were felt to be not dead radiation centers of physical forces, but the habitations of the gods.) This is beautifully expressed in the myth in which Aphrodite, arising from the sea (the waters of the universe), sets foot on land in Cyprus, the copper island. Cyprus was the richest source of copper in antiquity, and the name of copper, *cuprum*, may well have been taken from the name of the island. The medieval painter representing the Cyprian goddess being carried to land upon a conch shell, did not know, of course, that the conch is a mussel that breathes with copper, but nevertheless he painted wholly in the spirit of the ancient conception. Where the copper goddess lands is just where copper, out of the etheric cosmic streams, has so strongly condensed as an earthly ore.

The age of the intellectual and consciousness souls succeeded the sentient soul age. Iron took over the dominant place that once was occupied by copper. The iron peoples of the North stepped into the role of the copper peoples of the Southeast. Copper receded into the artistic sphere, serving the foundryman, the sculptor, the engraver, the etcher, the instrument-maker, and the organ-maker. But still it passed daily through their hands in the form of metal coins, even though these are humble pennies.

All this was changed at one stroke when the nature of the sub-material world, electricity and magnetism, was explored, and its forces were revealed on a scale that had never been imagined. Now electricity produced its own copper, which was to pave its way over the earth. We have, thus far, considered iron and copper as cosmic gifts and have followed their interworkings in the world of organisms right up to man. Now a sub-material world breaks into earthly activity, incarnating in the ways of iron and copper. A dynamo is a combination of purest iron and purest copper, all for the purpose of bringing electrical energy into the world to an extent that gradually rivals the cosmic gifts of light, warmth, and cosmic rays. Iron in dynamos is mostly electrolytic iron, copper is always electrolytic copper. The forces of electricity separate these two metals from all the other substances that nature has joined to them, and what remains materially is really only the abstract substances, Fe and Cu. But not before a tremendously complicated force-complex of electric activity has taken part in the production of this "pure matter." Formerly, all copper was smelted from the ore by fire. Today, most of the copper is refined electrolytically.

Man will have to achieve a conscious grasp of these processes if his servants are not to become his masters. He is beginning to unlock the mysteries of the forces that lie enchanted in matter. These forces are also connected with the mysteries of his own bodily existence. They will be of lasting benefit to him only if their spiritual aspect is revealed to his own spirit. Then he will be able to use these forces in a truly creative way.

VIII

QUICKSILVER

When we study the distribution of metals through the earth, we find in quicksilver, as in others, regularities pointing to a great organism. Just as every organism is a composite of the activities within its various organs, so the earth contains metals everywhere, in all its parts and in every kingdom of nature; in tangible condensations, however, the metals, each in its own way, are concentrated in definite and significant localities. Quicksilver, more than any other, may be called a European metal. Central Spain (Almaden), Central Italy (Monte Amiata), Northern Italy (Idria in Carniola), and Southern Russia (Nikitovka) have always produced the lion's share (about 85% in the last decade) of the world's quicksilver. The mines of Almaden have been in operation since antiquity. Europe is far and away the richest continent in quicksilver, far surpassing all other deposits on the globe, such as those in Nevada, Texas, California, and Mexico.

Thus the zone of greatest importance for quicksilver is in a central position between East and West, North and South.

In these deposits quicksilver appears in a single ore form, that of cinnabar, the scarlet red sulphide of mercury. In this it strikes a definite note in its relation to sulphur. This relation hovers in the background when we find sulphides of silver, copper, zinc, and antimony together with quicksilver in the much rarer mercury-bearing fahl-ores (gray copper ores or tetrahedrites) and in living stonite. Here, too, the metal follows the ways of sulphur; it combines with silver and copper, and where quicksilver, silver,

119

and copper work together, antimony joins in. In contrast to these strong affinities for sulphur is quicksilver's extreme reluctance to combine with oxygen, carbon dioxide, water, and other processes that are so characteristic of iron, copper, lead, and tin. Quicksilver keeps to itself, avoiding the other important metal formations.

In addition to cinnabar we find the pure unmixed metal in nature as native mercury in the lower depths of the cinnabar deposits, interspersing the rock with the finest drops, but only in meager quantities. It stands before us here like a wonder of nature, the only liquid metal on our planet, one of the "revealed mysteries" of our earthly world.

CINNABAR.

We must pursue the ways of sulphur in nature when considering the quicksilver minerals, above all the most important one, cinnabar (HgS), which may justly be called "quicksilver blende." The natural metal sulphides divide into pyrites, glances, and blendes. The hard lustrous pyrites still show clearly the metal quality of their nature. This is partly subdued in the softer glances, and is quite overcome by sulphur in the soft, glassy, translucent blendes. The beautiful fiery-red cinnabar crystals no longer show anything metallic, but rather remind us of semi-precious stones. A special nature-process, this meeting of sulphur and mercury! The scarlet-red color manifesting here must be regarded as a peculiar phenomenon, for in its various salts and combinations quicksilver is not otherwise a colored metal, such as chromium, copper, or nickel. Only in the presence of sulphur does it rise to such a "mineral inflammation." In this inflammation sulphur loses its volatility and quicksilver its liquidity. It now manifests color alone, but this it does vigorously in the dense, heavy, passionate red that seems to bore violently into our eyes. But if, by looking fixedly for a while at cinnabar against a grey surface, we try to find the color by which the eye seeks to balance out this yellowish-red, we will experience a forceful blue-green. This is the color of the glowing metallic vapor of the mercury lamp; the two colors are complementary to each other. Sulphur brings the metal into a condition where it is no longer characterized by lustre and great weight (shutting out the light, embracing the darkness, and submitting to earthly gravity), but becomes pure color. This color is the

120

most energetic intensification of the active side of the spectrum (from yellow to yellowish-red) that the mineral nature offers in the way of pigment. Expressed in the sense of Goethe's theory of color, yellow, the primal phenomenon born out of the light, has, by the activity of darkness, become intensified to its highest degree, without itself becoming creative as a color, for this yellow-red entirely lacks the violet or blue tones. Cinnabar red is the color preferred by strong, healthy, but primitive natures, the color least avoided by children starting to paint, as Goethe points out in his chapter on the moral effects of color.

QUICKSILVER AS A METAL.

The pure metallic condition is entirely suited to the nature of quicksilver. The natural processes themselves bring about this condition, so that quicksilver manifests itself in an almost noble steadiness. If quicksilver were solid, it could be ranked alongside silver as far as its outer appearance is concerned. It easily frees itself from its combinations. The ancients already knew how to produce it by grinding cinnabar with vinegar in a bronze mortar (Kallias, 415 B.C.).

THE "METALLIC WATER."

It is one of the great miracles of nature that quicksilver, although almost twice as dense as iron, fourteen times heavier than water, and having one of the highest atomic weights, is nevertheless liquid. This is the leading quality of quicksilver, from which much else will become comprehensible. At the slightest nudge it disperses into drops and tiny droplets. But then it comes together again just as easily. Its power of cohesion is extremely great. It permits each drop to round itself out and close itself off. It causes quicksilver in a narrow tube that has been dipped into a larger vessel to remain at a lower level than the one in the surrounding vessel. On the other hand, its adhesion to its surroundings is slight. It does not moisten what it touches but, after first scattering into countless little drops, immediately and completely returns into itself.

It would, however, behave differently if its surroundings were composed of metals. It would moisten them, give itself over to them, develop powers of adhesion. Thus, in many respects, quicksilver is for the world of metals what water is for the earth. It dissolves metals as water dissolves salts. It dissolves above all the plastic, flexible, soft, "inwardly liquid" metals, such as gold, silver, tin, lead, copper, zinc, cadmium, or the alkaline metals; it absorbs them, and forms amalgams with them. Less soluble are the brittle metals of the large-crystal formations, such as antimony, bismuth, and arsenic. But iron and its relatives, such as cobalt and nickel, resist quicksilver. Thus it is possible to store and ship quicksilver in iron flasks. Aluminum, too, is disinclined towards quicksilver's power of solution, but we know that iron and aluminum enter into the earth-process in a wholly different and much more intense way than the other metals.

In the youthful days when the earth was more life-permeated, softer, less solid, all metals were once liquid. Its aging was accompanied by processes of stiffening and rigidification, which invaded the metallic nature also. Quicksilver alone did not succumb to them, but held fast to the earlier conditions. It remained "young," as it were, retaining the earlier cosmic forms of the earth's first days. Not until subjected to the cold of polar winter, minus 39° C., does it solidify to a silvery, flexible, soft, malleable, and ductile metal. This proves to be a good conductor of heat and electricity, thus ranking itself with copper and silver in contrast to lead, tin, and iron.

Still earlier earth conditions were gaseous, vaporlike. Quicksilver has retained a resemblance even to these conditions, being not only liquid but also the most volatile of metals. It boils at only 359° C., and even at normal temperatures it volatilizes so strongly that the small quantity contained in an ordinary thermometer can easily fill the air of a large room with vapor.

QUICKSILVER LIGHT.

The fine glowing vapor in the mercury vapor lamp, radiates an intense greenish-bluish light, which takes away all the effects of the "warm" reddish colors. In this light the human skin assumes a gray and corpse-like pallor. This light, however, is rich in violet and ultra-violet rays, thus bearing a certain resemblance to the natural light of great altitudes, so that with its help the so-called "alpine sun" lamp is produced. This light is cold, but chemically effective. This is one of the peculiarities of quicksilver, along with those of the solid or liquid conditions. Quicksilver light, with all its properties, is still quicksilver. Only it is quicksilver in light form. The strong chemical effect of this light is one of its most prominent attributes.

The chemical forces in the world appear and work primarily in the liquid realm. The fact that quicksilver remains liquid in spite of its weight and density indicates that it has a special affinity for the forces that are active and at home in the liquid, i.e., the field of chemical activity. This becomes immediately apparent when, by the device of the mercury vapor lamp, the coarse materiality of this metal is torn out of its space limitations and, as it were, inwardly "cracked," so that it streams out in light-form into the limitless. Then chemical activity begins to radiate.

With its help we can accomplish interesting chemical changes. Ergosterol, for example, when irradiated, can be changed into a substance resembling Vitamin D.

CATALYTIC EFFECTS.

The coarsely material condition can be overcome in still another way by forming the largest possible surface, as is achieved by extreme thinning. Here the materiality strains to get out of the three-dimensional and into a two-dimensional state, which comes closer to the nature of light, and lifts the matter out of the realm of gravity. In this condition quicksilver is an important catalytic element, which makes possible or accelerates the combination of certain substances, again a distinctive form of chemical activity. Quicksilver acts here as a uniting rhythmical element, harmonizing the chemical polarities.*

CHEMICAL CHARACTERISTICS.

We have already mentioned that the metal nature of quicksilver rises almost to noble rank. It is impervious to the influence of its surroundings such as dampness or air; it does not readily enter into any relationship with

---

* This is a real "mercurial" property. It is also possessed by other metals insofar as they are mercurial in the sense mentioned in Chapter 6.

earthly matter. Compare this with iron, to appreciate the full contrast. Under ordinary conditions, for example, quicksilver is never affected by oxygen; it does not rust. But when heated close to the boiling point and just about to vaporize, it suddenly begins to work with the oxygen. This transforms the metal into a reddish-yellow oxide of mercury. With still further heating, however, the oxygen is again released. We can say that quicksilver "breathes" oxygen in or out according to the degree of heat, thus acting like a small metallic lung. (With this method it is possible to produce pure oxygen from air.)

In its compounds quicksilver occurs in two valences. In the mercurous compounds it is uni-valent, like silver, in forming a light-sensitive, insoluble chloride that, in the form of mercurial horn ore is a rare mineral in nature (as is silver chloride) and has to be artificially produced. It is known as calomel. In the bi-valent state it appears in the mercuric compounds, such as the corrosive sublimates and mercuric chloride. In its uni-valent condition quicksilver thus inclines toward the nature of silver.

For a heavy metal, its relations to carbon are unusual. It can combine with it directly and thus find entry into the realm of carbon chemistry, as shown by mercury trioxy-acetic acid and many similar combinations. Cinnabar, too, though not found together with other metal ores, is occasionally found in association with curious bituminous hydrocarbons, such as idrialin, in hepatic cinnabar. A fact that is like a metamorphosis of the carbon relation is that quicksilver, once it has been absorbed by organic tissue, is held fast by it and escapes only with great difficulty.

QUICKSILVER IN CIVILIZATION.

In our descriptions so far, quicksilver's character has emerged from the way it appears in nature. To this "natural destiny" we must add a description of its lot in the human domain. It enters this sphere not only by way of physiology, as a remedy perhaps, but also by rising into the domain of culture. It is part of the history of human development.

The oldest known use appears to be a cultic-magic one. In Kurma, Egypt, a small vessel containing quicksilver as a gift to the dead was found in a tomb of the second millennium B.C. A "Hermetic" mystery! Hermes-Mercury or Thoth, the god with whom this metal (and the planet) was associated, was the "Psychopompos," the guide who led the departed souls into the realm of spirits.

The pure metal was first described by Theophrastus in the third century B.C. In the first century B.C. Vitruvius describes how it can be extracted by distillation. Here the metal moves gradually out of nature into the social sphere.

At first, the forces of this "metallic water" are utilized in connection with the precious metals. These appear mostly in fine distribution, only a small portion being concentrated in tangible quantities. With the help of the metal that had remained liquid from the beginning, the ancients learned to leach the minutest particles of gold and silver out of the ore, and to concentrate these in profitable quantities. The Romans transported considerable quantities of cinnabar from Spain to Rome, and extracted the metal from these. The Moors discovered the quicksilver alloys, the amalgams. The Spaniards developed this aspect of mining to such a point that, after the rich ores of the South American silver mines had been worked out, for a long time they shipped about a hundred tons of quicksilver per year to the colonies for the purpose of exploiting the poorer ores. All this quicksilver came from the mines of Almaden, which must thus be reckoned the oldest in the world, and almost the only ones that have been successfully worked from antiquity to the present day.

The fates of the individual metals in human culture are diverse. Gold and silver are handed to the artists, who fashion beautiful and imperishable cultic objects from them, which they then turn over to the priests. Mercury, however, falls to the merchants. For over a hundred years the family of the Fuggers controlled the quicksilver trade of the entire world. In modern times the world trade was again monopolized by one powerful family, the Rothschilds.

Another "Hermetic" mystery of quicksilver was guarded in the laboratories of the medieval alchemists. With its help they sought to find ways to transmute matter, parallel to which the soul of the adept would be transformed and raised. They still saw in quicksilver something that was, in a higher sense, capable of transformation, something that had not yet come to a dead end, as had the other metals.

On its next level of activity, beginning in modern times, quicksilver is used as a remedy to an ever increasing degree. Here it becomes strikingly clear how the bright mystery of healing has behind it the dark one of poison. When syphilis became so dreadfully prevalent in the 16th and 17th centuries, mercury was the most important remedy. It cured the disease, but created a new and graver malady, thus revealing the poisonous side of its nature. This has been known only to the miners who, in digging this ore, often fell victim at an early age to the lingering death caused by its metallic vapor.

Next it serves to show man his physical image. The mirrors of the Middle Ages, as well as those of modern times until half a century ago, consisted of glass coated with an amalgam of tin and quicksilver. The shadow side of this achievement was the mercury poisoning of the mirror-workers.

Now mercury becomes the metal of the physician in more and more fields. The auxiliary tools such as the thermometer and the blood-pressure apparatus, the sublimate used as a disinfectant, the many remedies in the treatment of syphilis, dropsy, etc., the grey mercurial ointments, the precipitate ointments of the ophthalmologists, the amalgams of dentistry, all show the wide field of application, where we can also observe many harmful effects proceeding from even the smallest traces of quicksilver. Entirely new areas of efficacy were discovered by the homeopaths who, by happy management of the dosage, were able to avoid the sombre experiences with their poisonous side-effects.

Quicksilver also enters the realm of modern technology. Quicksilver air-reduction pumps make it possible to produce high vacuums and all the light phenomena resulting from the "de-airing" of matter, on which techniques of irradiation—again a medicinal procedure—are based. A modern laboratory is unthinkable without the considerable quantities of quicksilver that go into its switches, pumps, and measuring apparatus of a chemical and electrical nature.

So we have the "deeds and sufferings" of quicksilver in the human realm. It is created out of natural forces from the macrocosm. It is transformed by man. It is still liquid now; some day it will become solid. Thus Rudolf Steiner describes it. It will then have properties that will express what mankind has achieved morally up to that time.

QUICKSILVER AND THE ORGANIC WORLD.

Iron alone among the metals is absorbed by the human organism in notable quantities. Copper, lead, and numerous other metals are also found in the body, but only in minute traces. Obviously, the important thing is not the *material* nature, but certain *activities* that begin their manifestations as the materiality recedes and becomes inessential. (In the inorganic we have similar activities in the catalytic processes.)

"The organism is a complex of activities." * Hence, it looks to the substances not primarily for their material nature, but for the stimulus to action that they contain, each in a different way.

There now arises the question as to what activities in the organism can result from the activities slumbering in the quicksilver substance.

The human organism is, like any other, a totality that stands on its own feet, but it is one that has manifold members. To begin with, it is differentiated into highly diverse organs. Each organ is not so much built up of cells as articulated into cells. The aspect that makes the organism into a whole, and the one that separates it down to individual cells, may be looked upon as polarities. The organs, which are less than a totality but still more than cells, stand in the center.

Every organ, every part of an organism, must be attuned to the totality. It must have a life of its own, but this must not be a *self-willed* life. We have learned how to keep cells—even entire organs—alive in nutritive solutions when entirely cut off from the rest of the organism. This shows that cells and organs are capable of a life of their own. But if they live outside of the organism, they can do so only *parasitically*, at the expense of some other foreign totality. They then develop the tendency to destroy that totality. They are "artificial parasites."

A parasite lives without concern for the totality from which it draws its existence. It can only destroy a totality, never build it up. Since many parasitical creatures are themselves merely heaps of cells, not genuine totalities, parasites can live only where a wholeness disintegrates. (This is

---

* "The essence of a living organism lies not in its substance, but in its action. The organization is not a system of substances, it is activity." R. Steiner and I. Wegman, *Fundamentals of Therapy*. All the following quotations in this chapter are from this book or from *Spiritual Science and Medicine* and other lectures given to physicians by Rudolf Steiner.

why it is unbiological to begin the study of biology with an examination of the unicellular beings.)

The single cell must never have too much life; there must be no super-life. A process must be active in the organism that continually deprives the cell of this super-life, seeing to it that the part remains integrated into the whole and that the whole maintains its government over all the parts.

Among the totalities known to us, the one on the grandest scale is the cosmos. In the self-willed life of a cell we find the polar opposite of this totality. The spiritual investigator, who has an inner vision of this cosmic life as a totality, describes a process that is of great significance here. "The cell develops a wilful, perverse life. This is opposed by something outward that deprives it of life, giving it the form of a drop. In the drop-form the result lives between something that is striving toward life and something that is sucking this life out of it. This is the mercurial nature, which strives obstinately to become a living drop, but is drained by Mercury of this super-life. Every drop of mercury would be alive, if it were not for the planet Mercury. The organs within us that are most strongly inclined toward the cellular, the abdominal organs lying between the principal excretory organs and the heart, are those with the duty of keeping the cellular in order and, by means of Mercury, preventing its proliferation. Thus these organs have their relation to quicksilver."

There are two processes latent in quicksilver; one that strains toward life, and another—the cosmic Mercury process—that sucks this life away. But the same processes are active in man, except that he brings them about by his own nature rather than through material quicksilver.

The first process becomes visible in man when we follow the nutritive process. In the first stage of digestion, the food is broken down into inorganic material. Having passed through the stomach and intestines, it becomes almost mineral. The nature of the inorganic is that one particle lies beside another without any higher coordinating totality—an atomized chaos, as it were.

What has thus been cut to pieces and dissolved into its parts must now be taken hold of by the formative-force organization and fitted into the living whole of the organism. Here we have a region of vivifying and healing. For the food is, at first, completely foreign to the organism, hence a poison. The process of nutrition must be arranged so that the poisoning, as it arises, is immediately overcome by a healing counter-process. (That foodstuffs are poison for the organism is shown by the anaphylactic shock that results when protein is injected directly into the bloodstream rather than going through the stomach and intestines. In the normal digestion of many people, certain foods prove to be too strong for complete "detoxification" by the organism, whence "allergic" diseases may arise.)

Now we come to the second process. What has been atomized must be enlivened. But the enlivening must not take place "atomistically." The parts must not be too lively or too independent. There must be no super-life.

This must begin in the intestines with the intestinal flora and fauna. These are deprived of liveliness to such a point that no such processes arise as we find in the flora and fauna in outer nature. Only a muted life can be allowed in the intestinal region. It is outside the intestines that the atomized food is enlivened and, in the form of droplets, taken into the lymph and blood. As it gradually rises into the organism, it enters the sphere of lower organs that remain as strongly cellular as, for instance, the liver.

The entire area sketched here is a field where quicksilver is useful as a remedy. Inside the walls of the intestines, it works on general nutritional difficulties, infectious diseases of the intestines (typhus, dysentery, infantile diarrhea, even intestinal tuberculosis). If atomistic organization occurs in the field of mercurial processes outside these walls, a disposition to syphilis may arise. "When the digested matter has passed through the intestines, it comes under the influence of cosmic rounding-off forces. In these the ego organization is primarily active. Reaching too strongly into the metabolic system, it would be the tendency of the ego organization to round off, to organize, individual members instead of organizing the entire body in its proper form. All forms of syphilis are the result of such atomistic organizing. The ego organization reaches into small systems of the human organism,* which properly should be left to the organizing forces of the etheric. When quicksilver is administered to the organism, it imitates, more strongly than any other earthly substance, the outer form of the cosmos. The atomization is given over to the quicksilver, and the ego thereby becomes free again."

---

* The processes in the cells must be intended here.

In the extra-human world we find this power, so hostile to atomization, in the strong toxicity of quicksilver compounds toward one-celled parasites. This explains the use of mercury in anti-fungus baths for seeds, and of corrosive sublimates as disinfectants.

Quicksilver also affects the glands. Mercury poisoning induces excessive glandular activity, such as salivation. Calomel strongly stimulates the activity of the liver. The glands are organs that, in their form, do not rise appreciably beyond the cellular. Their task is to fill themselves with their respective secretions. These secretions are then rhythmically pressed out by the intervention of the supreme totality of the organism. It is the astral organization that brings this about, but the activity can be usurped by quicksilver.

But quicksilver does not only oppose any declaration of independence by the cells. Its tendency to reintegrate any separatist existence into the whole applies to *all* processes that try to set up their own shops within the organism. Quicksilver brings these detached forces back into resorption. Wherever isolating processes arise in the organism, Mercury leads them back into the fold. Examples of this are the catarrhal processes. "They arise because some tract of the organism has, by outer influence, been wrenched out of the control of the organism as a whole. This is the case in catarrh of the bronchi and adjacent areas. If the forces of quicksilver are applied here, they exert a healing effect." By combinations with sulphur, "which is effective in the region where circulation and breathing adjoin," we carry the influence of quicksilver into the lungs and neighboring parts.

"We can use quicksilver therapeutically wherever separatist processes arise that must be brought back under the governance of the whole organism." Organs that are open to the outer world, such as the senses (which sit almost parasitically upon the organism), the skin, or the breathing organs, are always as it were, lifted out of the organism. They incline toward independence. Thus they are especially subject to colds, for example. If they are unable to overcome, to "digest," these invasions of cold, an alien element will displace their own warmth organization. As a counter-process we have inflammations, which increase the heat around the organs in question. If the region is not brought under normal control, if the "alien organization" of the cold invasion is not expelled, there arises the tendency to eject this alien element. It is threatened by decomposition. The result

130

may be pus formations. In such cases one can use quicksilver as a remedy. This is the basis of the healing effects of yellow precipitate ointment in inflammations of the conjunctiva and the cornea; of sublimate in chronic tonsillitis, and (used externally) in otitis externa and media, as well as in panaritium (whitlows). Cinnabar has proved its efficacy in acute and chronic inflammations of the tonsils, likewise in abscesses of the teeth. Then there is the use of mercurius solubilis by the homeopaths in inflammations of the mucous membranes, the glands, and tissues, when pus formation is feared.

We should add that "quicksilver is the congealed process that stands right in the middle of the propagation processes that completely separate the organism from its own nature." This explains its effectiveness against syphilis, and the insensitivity of the syphilitic patient to the obvious damage from the poison of mercury.

The muting of the independent action of the cell processes by quicksilver explains its use, for example, against breast sarcoma in connection with viscum. Here the metal should be in colloidal form, i.e., an especially refined form of the drop-condition.

A third area of healing influence is that of certain rheumatic illnesses, such as chronic muscular rheumatism, acute rheumatism of the joints, rheumatic fever, and arthritic swelling. "Metabolism not completed by the ego manifests as rheumatism." Whenever the organism takes up nutriment without divesting it in the digestion of all its own forces, this nutriment is likely to assert its foreign notions without any deference to the totality, which is supposed to be ruled by the ego. We then have allergic diseases, tendencies to eczemas, to uric conditions, etc. The liver activity may be too weak, as a result of which rheumatism of the muscles and joints may arise. As far as these disturbances originate in the incorrect or incomplete transition from the outer to the inner digestive process, where the lymph and blood are formed (and where the effects of copper are known), it will be possible to use mercury as a healing agent.

IX

SILVER

Silver, always deemed the most precious metal next to gold, is rarely found in a pure and unmixed state. The dainty shapes, moss-like or fern-like, in which it thus appears are truly "art-forms of nature." Pure silver ores are also rare. These appear as sulphides, especially silver glance (argentite); as antimony-bearing ores, such as pyrargyrite, silver-antimony-glance, and the brittle sulphurous ore known as stephanite; as arsenic-and-antimony bearing ores; as the fahl ores, tetrahedrites, which contain copper, silver, iron, zinc, quicksilver, antimony, and sulphur.

Thus silver on the whole, lacks the ability to form any sizable deposits of its own.* But, characteristically, it occurs together with gold, lead, and copper, which form extensive deposits. Gold deposits frequently have considerable admixtures of silver. All lead glance contains silver, and thus is the most important silver ore. The important copper ores of a sulphurous nature contain silver.

In the distribution of this metal over the globe, we find Mexico to be the richest silver country, then the United States, South America (Peru and Chile), and Canada. The deposits in the rest of the world fall far behind

---

* Among the deposits in which silver ores do predominate are those in Freiberg, Saxony; St. Andreasberg in the Harz Mountains; Schemnitz in Slovakia; Kongsberg in Norway; and others in the Black Forest, South America, and Mexico. But most of these are exhausted. Those still worked yield only a fraction of the world production of silver. They bear sulphides, mostly silver glance, in their deeper strata, and much pure silver in the upper oxidation-zone.

these. During the Middle Ages, the mountainous regions of Middle Europe, Sweden, the Harz, Bohemia, and Hungary yielded most of the silver. During antiquity, it was Spain.

A clue to the distribution of silver may be obtained from the following figures for silver production in 1931:

Europe	459 tons—about	8%
Asia	417 " — "	7%
Africa	48 " — "	1%
America	4878 " — "	80%
Australia	268 " — "	4%

Most silver, however, occurs in sea-water, although in an attenuated solution.

Although most of the silver found on solid land is concentrated in the west, in the new world, once it has been won by the hand of man it has always followed a curious course toward the east. The silver mined in Spain during antiquity wandered toward the Orient; the Phoenicians grew rich trading in it. To this day, the peoples most attracted to this metal are those of India and China. Their money is based on a silver standard. In contrast, the bulk of the world's yield of gold streams toward the west. America hoards most of the gold, though the deposits are largely in Africa. In the social sphere, gold follows the course of the sun, which sets in the west. Silver flows to the east, opposite to the course of gold.

To begin with, let us look at the pure metal. In its interaction with the substances and forces of the outer world, it reveals itself to us in phenomena that speak eloquently of its nature.

Silver and Light: The Perfect Mirror.

In the cool and noble lustre of the pure metal there is perfect mirroring power. No other metal equals it in this. It returns the instreaming light undimmed and almost unchanged, declining to absorb any of it for itself. Nothing of its own nature is mixed with this light; neither color, as with gold or copper, nor turbidity, as with lead, antimony, etc. Look into a silver mirror and you see nothing but mirror-*pictures;* the silver itself withdraws completely. Today's mirrors, therefore, are made almost exclusively by coating glass with a fine layer of silver.

In the gradual formation of such mirror layers by vaporizing silver in the vacuum of a quartz tube, there appear on the walls of the tube, breath-thin transparent coats of the most magnificent colors: rose-red, then red-violet, then blue-violet, and finally a wonderful blue. To appreciate this color sequence, let us look through a prism at the fine pointed end of a black band on a white background, with the edge of the prism parallel to its longitudinal axis. Darkness will dissolve to color in the light; at the tip it will be purple, changing gradually into violet and blue on the side where the darkness shines through the brightness. The pure purple first appears where the light is in active balance with the delicate film of precipitated silver. With gradual densification, the color takes on more and more of the blue that manifests out of the "power of darkness" of matter. Finally, the metal layer becomes opaque and begins to reflect.

THE LIGHT-SENSITIVE METAL.

Despite its precious nature, silver is easily dissolved by acids. The resulting silver salts have, in certain of their qualities, a remarkable resemblance to the salts of the alkali metals (about which more will be said later). Silver might therefore be termed an alkali metal that has become noble. But only darkness leaves it in this salt condition. Light immediately gives it the tendency to revert to the metallic condition. This extraordinary sensitivity to light is, of course, the basis of photography. Primarily it is silver chloride, silver bromide, and silver iodide that are so light-sensitive. But precisely here lies the difference from the corresponding alkali compounds. Silver chloride, for example, in contrast to sodium chloride (rock salt), is not water soluble; in nature it occurs not as a crystallized mineral, as does sodium chloride, but as a horn-like substance. It is actually called "horn silver" (cerargyrite).

THE MOST COLOR-SENSITIVE METAL.

When silver chloride, produced as a fine layer upon a silver plate, is put under light of various colors, it assumes the color of the light to which it has been exposed. It turns red under red light, green under green light, blue under blue light. In so doing, it changes from its salt condition into a finely distributed colloidal condition. Thus, to a higher degree than any other metal, it is an image of the interaction between light and darkness that actually brings the colors to birth. (Since this phenomenon is fleeting and changeable, it cannot be fixed and therefore is not a useful basis for color photography.)

THE BEST HEAT CONDUCTOR.

Of all the metals, silver is the most pervious to heat. In metallic vessels of equal size, equally heated, water boils first in one of silver, then in one of copper, and then in one of gold. It boils more slowly in iron, tin, and lead. Silver yields most freely to the passage of heat, retaining little for itself. Heat expands into space fastest by way of silver. A leaden rod can melt at one end while the other is still cool. Enormous heat differentials can occur in such a rod, but not in one of silver. Heat imparted to any portion of a silver mass is immediately spread through the rest of it.

SILVER AND ELECTRICITY.

Silver is the best conductor of electric current. In this respect it is related to the "good conductors," gold and copper, whereas iron, tin, and lead are poor conductors. The condition of electric tension, which is the prime phenomenon in electricity, is readily equalized through a silver wire; silver, we can say, causes the tension to disappear most quickly.

Most metals achieve a high conductivity when they are cooled to low temperatures. Silver does not need to be cooled in order to conduct well. More than any other metal, it behaves as though "inwardly cold" at ordinary temperature.

SILVER AND MECHANICAL FORCES.

When facing mechanical influences from without, silver proves to be soft, ductile, and flexible. It can be drawn into fine wire, and is surpassed only by gold in its pliancy and cohesion. (One ounce of silver may be drawn into a wire almost thirty-five miles long.) It does not vigorously assert its own form through hardness and rigidity, in the manner of the "brittle metals" that show a crystalline structure, such as antimony. Natural pure silver prefers, as mentioned before, plant-like rather than mineral-crystalline forms. The extraordinary pliancy, the ready adaptability of this metal, bring into the world of solidity something not native to the solid state, that is, a metamorphosis of the forces that rule in the liquid condition. Silver readily abandons the solid condition, as shown by its comparatively low melting point. It also vaporizes rather easily. Only when, out of the salt solution, it is converted to a metallic state by electric current, does it create crystals that are rich in form. Its shape then betrays the form-forces exerted by electrical coercion.

In spite of this plasticity, however, the structure of silver has an inward nobility that manifests in its pure ring. Silver flutes and silver bells sound especially pure and clear. Coins with a high silver content have a good ring. This is unusual, since soft metals generally have a poor tone, like lead. They have to be rapidly cooled (by immersing them in liquid air, for example, which makes them contract and solidify) if they are to have a good tone. Silver does not keep the sound to itself, but permits it to go forth freely as tone, changing it as little as possible by its own nature.

Something plays into the soft substance of silver that reveals itself both as a cold activity and a formative power. This is a mysterious contradiction.

Chemistry of Silver.

In the periodic system of the elements, silver stands in the same column as the alkali metals. It resembles these more than any other heavy metal. It is chemically uni-valent, like the alkali metals. Some of its salts crystallize in shapes like these, even forming mixed crystals. Hydroxide of silver is strongly alkaline, somewhat like sodium hydroxide or potassium hydroxide. The alkali metals are halogens, salt-formers. Sodium chloride is salt par excellence. Its original home is the ocean, which also contains most of the silver, though in high dilution. The metal sodium can be produced from common salt only by highly artificial and drastic processes, and it has only a short span of life since it strains with all its might to get back to the salt state to which it belongs. Its metal state is really "stunted." Silver, however, finds its natural condition in the metal form. It is, as we said before, an alkali metal become noble.

Silver stands in a peculiar relationship to air. Melted silver greedily sucks in the oxygen from the air—up to twenty times its own volume. It lives in close unity with it as long as it remains liquid. But as soon as silver solidifies, it suddenly loses this retentive ability, and expels the oxygen with explosive violence. The metallurgist calls this phenomenon spattering or splashing (German "Spratzen"). Hence the surface, which in the liquid state was so beautifully smooth and lustrous, hardens into formations like moon-craters or pockmarks, because the bubbles of gas break loose at the moment of solidifying.

As a solidified metal, however, silver remains unalterably "noble" in the face of water as well as air. It would remain always bright and shining were it not so sensitive to *sulphur,* the scantiest traces of which mar, discolor, and blacken its beautiful lustre. Relations to sulphur become apparent here that we have already met in the most important silver ores, but now in another aspect.

PHENOMENA IN THE ORGANIC REALM.

In the behavior of a substance toward its surroundings there emerge what we call its properties. In this interaction, however, the substance reveals two things.

First, what it rays into the surroundings as its own nature, what it modifies the surroundings with. The effect of this decreases with distance. This indicates to what extent a substance exists in and of itself. In these properties the lifeless mineral world speaks to us.

Second, how a substance is taken hold of by its surroundings. Through such properties the substance combines with its surroundings into a whole. This whole, when sufficiently comprehensive, we call the cosmos. These properties alone express the kind of role a substance can play in the organic world. Here we see effects that do not decrease with distance but are, in fact, determined by the surroundings. We may call the first effects central, the second peripheral. Central effects must be ascribed to central forces, peripheral effects to peripheral forces.*

Healing effects depend upon the manner in which a substance lets itself

---

* See R. Steiner and I. Wegman, *Fundamentals of Therapy.* All following quotations in this chapter, unless otherwise noted, are from this book.

be taken hold of by the whole of the human organism. What counts is not what a substance *is,* but what the organism can *do* with it.

What phenomena are displayed by silver in the life process? The activities stimulated by it are in harmony with the *up-building* processes. This is shown by its not being poisonous to the more highly organized beings, and especially by its not causing chronic poisoning. In contrast to lead silver jewelry may be worn on the bare skin without hesitation. Only the low parasitical "pathological" bacteria that live in darkness find silver hostile. In the Katadyn process (contact with silver-coated sand or porcelain rings), drinking water is freed from germs by contact with highly diluted silver. Surgeons confidently insert silver wires, plates, and pins into the body.

On the other hand, people whose occupations necessitate daily contact with lead, such as typesetters, are in danger of chronic lead poisoning. Association with lead arouses an abnormal degenerative activity, leading eventually to complete hardening and paralysis. As an antidote to these harmful lead effects Rudolf Steiner recommended foot baths in a diluted silver nitrate solution. Thus we have in silver an answer to the lead process.

But we have already seen that in extra-human nature, the important lead ores carry slight admixtures of silver. Thus nature has seen to it that lead shall not exist without an opposing force.

"Lead works upon the organism so as to increase the disintegrating action of the ego-organization." The consciousness processes are connected with this. They are bound up with the nerve-sense system, hence predominantly with the upper organization.

The effect of silver is diametrically opposed to this. The "homeopathic test picture" shows silver as follows. The limbs begin to obey the will reluctantly. There is a feeble response to the impulse to move. There is aversion to any activity, showing that the soul-spiritual has been forced out of the physical-corporeal in this area. Accompanying this is a dimming of consciousness, which may result in an unconquerable desire for sleep. A characteristic shift follows in the blood activity, which is no longer kept within its proper upper limits. The blood organ moves towards the head. The sense organs overfill with blood, the gums bleed, the teeth become loose. The head feels bloated and distended. Thus the activity of the

138

lower organization is driven upwards, where it works in polaric opposition to the normal disintegrating, shaping, consciousness-engendering tendency of the upper organization. In their proper place, in the metabolic-limb organization, these activities work constructively, although with a consciousness diminished to the point of perpetual deep sleep when compared with normal consciousness. The silver effects usurp the place of these upbuilding processes and drive them into the upper organization. In illnesses caused by a predominance of these processes, silver will bridle their excess.

Rudolf Steiner describes the silver effect thus: "Everything tending in the direction of silver is in a certain sense, as a metal, contrary to whatever tends more in the direction of lead . . . I say that silver and lead are polaric because silver works directly on the metabolic-limb system, and does this peripherally, working on what is external in the limb-metabolic organism, whereas lead works upon all that the head-organization has externalized. Silver stimulates the nerve-sense activity in the metabolic-limb system, and from there promotes the activity that permeates the entire body and stimulates the breathing in everything that yesterday I called the metamorphosis of the central heart-organ." (In another place: ". . . to which, in a pre-eminent degree, belongs the female uterus. The uterus is nothing else but a transmuted heart".)

In this connection we must look at the nutrition-stimulating and conception-promoting effects of silver. In the entire realm of the "lower organization," the soul-spiritual is strongly tied to the body, and is given over to its up-building; it acts unconsciously therein. What manifests here above all is the physical-etheric activity. Ego and astral body are submerged in it.

The first phase of this up-building is prepared by the digestive process. The metals active here are potassium and sodium. "They can be taken hold of by the part of the ego organization that is active in the system of the intestines." In this same area the silver effect begins. We have already seen something of the relationship of silver to the alkaline metals, and we even termed it "an alkali metal turned noble." Now we note the fact that silver combines closely with sulphur. The latter, however, resides in protein. "Sulphur is at the bottom of the entire process taking place in the intake of proteinaceous food. Sulphur passes from its foreign etheric nature (in the food) through the condition of the inorganic, into the etheric activity of the human organism. It is thus seen to play a role in the absorption of protein into the sphere of the human etheric body." The silver effect, in part, clings to this progression of sulphur. But then it goes farther in the organism, reaching as far as the periphery. This is especially noticeable in the secretion of metallic silver in the dermal layer of the skin when overly strong doses of silver salts are taken over a long period. It causes the so-called "argyria" or heavy discoloration of the skin. (Sulphur works as far as the skin, too, and silver is following it here "peripherally" in the most literal sense.)

Silver's stimulation of the nerve-sense activity of the metabolic-limb system directs our attention to these nerves, which are commonly called "motor" nerves, but really have the task, according to Rudolf Steiner, of perceiving. What is perceived, however, is not an outer world, but the inner world of metabolism. This perceiving remains, of course, deeply unconscious. When taken together with another remark made in a lecture in Stuttgart to the effect that the part of the nervous system that does not serve the uses of consciousness has the task of a sculptor with respect to the entire organism, this indication shows that silver's healing effects encourage not only upbuilding pure and simple, but a moulded and shaped upbuilding. Consequently, limp and spongy tissues are not only better nourished, but better formed. Furthermore, the blood process is strengthened with respect to the upper catabolic consciousness-engendering nerve-sense process; indeed, the latter is flooded with blood. Hence the blood is the scene of quite definite silver effects. We should remember here the virtue of high potency silver injections against fevers, especially septic fevers. "Such fever is actually an agent for a radical intervention of the disintegrating processes into the organism. We must immediately provide for a strengthening of the etheric body, so that this may paralyze the harmful effects of the astral body." Here highly potentized silver should be injected.

Thus we have the relation of silver to the abnormal organic warmth processes. It stops the generation of abnormal heat as compared to the normal temperature of the blood, and strengthens the etheric forces in the blood against the astral forces of disintegration. The alien processes, which occur as pus processes in the blood, are combated.

In the realm of the inorganic, we may get a hint of this ability of silver when, in its liquid, heat-permeated condition, we see it gulp up the oxygen that, in the organic realm, bears the life processes and transmits the etheric forces.

THE HISTORY OF SILVER.

The substances that confront us in the animate world are always something secondary, the results of something primary, of the formative processes. Substantiality arises in the living organism. To the extent that it can be set free and, thus detached, can lead a relatively independent existence, it must be said to have "died out of" the organism. Nevertheless, it will always have to be viewed in connection with the organism that formed it, if its properties are to be understood. The properties of the substance are, as it were, hieroglyphs, demanding that we read in them the deeds that have called them forth. A drop of oil of rosemary, for example, still speaks in all its properties of the warmth and light processes that, together with the nature of the rosemary plant, have created it.

The attribute or property of a substance, then, is its hieroglyph, or "high rune" as Novalis expressed it; it announces the deeds that created it. The deeds are in the past, but the hieroglyphs or properties still express the history of the substance and its creation.

In the animate world this is clear enough, since both the substance and the nature that forms it can be viewed together. But this can be accomplished for inanimate matter only if the entire earth is viewed as an organism, and not only that, but the earth together with all the forces of the universe that influence it.

* * *

Which of the formative events of the earth's history can be correlated with the substance of silver? Its characteristic properties speak clearly enough, but the ordinary history of the earth yields no facts that are related to these properties. It is different, however, with the results of the spiritual researcher.

In *Spiritual Science and Medicine*, Rudolf Steiner explains how "in the world of metals we must see the results of extra-tellurian forces and influences." Regarding silver, he points especially towards the relation between "everything of a silver nature and the undisturbed effects of the moon." Undisturbed effects of the moon, however, were possible only after our satellite became independent, i.e., after its separation from the earth. We can study the spiritual investigator's descriptions of the processes connected with this event in their relation to silver, and ask ourselves whether these processes betray something of the nature of this metal.

According to these descriptions of Rudolf Steiner, which can only be sketched here, after the previous separation of the sun, the earth (which still contained the moon) was left behind as a steadily darkening body since the inner light had been taken from it with the sun. Dark masses of water formed the nucleus of the earth, whose substance at that time had condensed as far as the liquid state. The formative forces of tone worked in the watery masses, and in these, in turn, worked the creative deeds of mighty beings. The dark and ancient waters of the earth's center slowly rose and were converted into dense fogs. The "water-earth" contained substances that were in solution or at most in a soft condition. The fogs became thinner with altitude, until the highest layers were translucent. The light working from outside transformed this outermost sphere into glowing vapor.

But the earth's evolution is unthinkable without man. During the union with the sun, man's body had reached the stage of the plant-like. Now he experienced a decline in his nature, a "clouding" of his being, from the darkness of the water-earth to which he had tied himself. He developed at this time to the stage of the amphibian. (He was, of course, never an animal in the present sense.) A part of his organization, the upper part, remained turned to the light. With this part man lived in the light-transfused vapor. With the lower part of his organization, however, he was related to the water that had been deprived of light. Through this he shared in a world of forces that were also dark and bad in a moral sense. His upper organization, however, in the etheric part that hovered above it, clung to the sun forces and thereby exerted a taming and ennobling influence upon the lower organization, which was permeated by fiery warmth processes from above. For the masses of vapor were pervious to heat, though not to light. In the myth of the dragon this stage of evolution is shown in a picture. The upper organization, endowed with the sun forces, was seen in the form of St. Michael or St. George defeating the dragon.

Tremendous purifying catastrophes now convulsed the earth. The waters condensed and a mineral nucleus arose. The moon, with its baser, evil, hardening forces, was expelled from the earth. The earth thereby became completely transformed. On the one hand, it experienced a "healthy" condensation of its core; on the other, the atmosphere became finer. It had been quite differently constituted before, but now it began to take on its present condition. Oxygen, for example, appeared for the first time. Now the light, the rays of the sun, could reach the earth again. Because it was disburdened of the hardening impulse, it underwent a rejuvenation and refreshing.

The physical moon, however, underwent a drastic solidification. It has ever since been wholly deprived of the fluid and the airy. Its scarred surface seems to preserve the blistered rigidity in which it was left as the air finally tore itself out of it in a way not unlike the spattering surface of solidifying silver that we described above. Since then, the moon has been the self-enclosed frigid world-mirror.

In connection with these earth processes the human form also went through a metamorphosis. The body, heretofore liquid, began to form bones, thus solidifying itself inwardly. The upper organization, however, changed in such a manner as to make it possible to breathe air.

Only one part of the moon nature was ejected to form the frigid, rigid, airless, fluidless body of slag that has been circling the earth ever since. Another part remained in the earth, with forces polarically opposed to the former. It works in the waters of the oceans where, by virtue of their silver content (according to Rudolf Steiner), it brings about the tides. But not only the fluids of the earth are subject to the tides; delicate inner "tides" in plants as well as animals show the earmarks of the moon in their rhythms. During full moon, seeds sprout more vigorously, the sap rises more strongly in the trees, the forces of reproduction are stimulated. We can only refer here to the many facts that have gradually appeared in the scientific literature. Thus we have something physical, the moon itself, that contains death and rigidity, and something etheric that has remained connected with the earth and has to do with germinating, sprouting, the vivification of liquids, and reproduction.

Now let us compare the properties of silver with the basic features of the separation of the moon that we have sketched. The facts that surrounded our satellite's hour of birth have set their seal on the nature of silver. The ring, the coolness, the mirroring faculty of solid silver; the way it changes on the departure of oxygen, which can combine with silver only when it is liquid and which afterwards leaves scars upon the surface of the silver all show that in solid silver we have an image of the part of the moon that had to leave the earth. Silver that has been dissolved to salts, on the other hand, points to an etheric moon sphere that has remained united with the earth, as is shown by its relation to the alkali metals that are so much at home in the sea, by its stability as a fluid when in the dark, as compared to its extreme sensitivity to light, and by its happy participation in what results as color from the interplay of darkness and light. We will realize this better if we do not emphasize the details but permit the whole to work upon us.

* * *

Thus silver takes its place in the cosmic context, so that we can call it, with full justification, the metal of the moon. But now we call it this out of our modern consciousness, without leaning on any old atavistic stages of consciousness which, in their dreamlike clairvoyance, saw these connections directly. The metals we have described up to now are a sevenfold challenge to us to expand our earth consciousness to a cosmic consciousness.

X

ZINC

An important feature of zinc can be brought out by pinpointing the localities on the globe where this metal is found in significant quantities. These localities coincide to a large extent with those in which lead occurs. Zinc and lead occur in common deposits as the sulphides, zinc blende (sphalerite) and lead glance (galena), arranged for the most part so that in the upper veins galena is either alone or predominant, while toward the depths it gradually diminishes as the zinc blende increases. Thus, paradoxically, the heavier lead ore lies above and the lighter zinc blende below. Another metal that is always present is lead's polar opposite, silver.

In nature, zinc follows the ways of sulphur, as do lead and silver. Zinc blende (zinc sulphide, ZnS, sphalerite) is by far the most important zinc ore, representing 90% of all deposits. Next is zinc carbonate or smithsonite, but this is already a secondary form, a transformation of zinc blende by atmospheric influences. In the carbonate deposits we find zinc, which always enjoys the company of iron, as carbonates, then as hydrates, and finally as oxides—the so-called flowers of zinc, corresponding to the "flowers of iron" (flos ferri or aragonite). Then there is zincite, which often occurs together with iron oxide, and finally there is zinc silicate, the silicious calamine or hemimorphite.

Nowhere in nature does zinc occur in the pure state. This must be brought about by man with the aid of carbon by reducing the ore (from which the sulphur has first been roasted off) and then distilling it in a retort. Then we have a bluish, rather soft, brittle metal of a leafy-crystalline structure, which melts somewhat later than lead, at 419° C., but vaporizes earlier, at 906° C. Its density is 6.9. Like all the brittle metals, it conducts heat and electricity only moderately well. Compared with a conductivity of 100 in the case of silver, its conductivity is 30. Thus this metal contains rather more of rigid form-force than of flexibility. When heated, it expands strongly, but it contracts just as strongly when cooled. Among the heavy metals, only lead resembles it in this respect.

Chemically, zinc is non-precious, combining readily with all earthly processes. Oxygen, water vapor, and carbon dioxide take hold of it with ease; it dissolves in every acid. The salts that it forms are largely water soluble. In these properties zinc resembles bi-valent iron and magnesium. In the periodic system of the elements, zinc falls into the same column as cadmium and quicksilver. Thus something in the nature of this metal points to quicksilver. But these connections are produced only by the abstract reasoning of the chemist; nature does not "obey" these relationships because it has not included zinc in the large quicksilver deposits. On the other hand, it permits zinc to occur together with lead, silver, and even iron—relationships that are incomprehensible to the chemist.

* * *

As shown in Chapter 3, it is the forces of cosmic radiation of the Saturn sphere that have endowed the earth with lead. Lead is due to the raying-in of the "undisturbed effects of Saturn." Since zinc is so strongly tied to the presence of lead, we naturally look for Saturn effects also with respect to zinc—not undisturbed effects now, however, but such as are combined with those of other planetary spheres. The Saturn process, as it extended more deeply into the earth, apparently lost its producing force for producing lead and gradually connected itself with something more closely related to the earth. A metal such as iron actually has a far stronger relation to the earth processes than the Saturnian lead.

Comparing the properties of zinc with the classical seven metals, we find it lacking in the colorfulness of copper and gold; in the light-sensitivity of the salts of silver, copper, quicksilver, and iron; in the flexibility and conductivity of silver, copper, and quicksilver. It is more related to the metals of the planets "above the sun" than to those "below the sun." We must place it along with lead, iron, and tin. Only its easy vaporizing reminds us of quicksilver. Let us draw no hasty conclusions, however, but merely contemplate the phenomena.

* * *

Zinc is essential to life in the plant world, even though only in diminutive amounts. Deficiency of zinc is evident in low chlorophyll formation (spotted leaves on fruit trees), in dwarfing (in tomatoes), in a stunting of the growth axis. The seed embryos of grain and of coniferous trees are relatively rich in zinc. In plants and segments of plants that are rich in sugar we find little zinc, but in those that are rich in protein there is much of it. Sugar beets contain as little as 2 parts per million, whereas mushrooms have as much as 280 per million. Some plants absorb an astonishing amount of zinc from soils that are rich in this metal, e.g., the calamine violet (*Viola calaminaria*), which concentrates it in its leaves and stems, but less so in its seeds. In its ash, we find about 1% of zinc. A higher zinc content, however, hinders the growth of most plants. The grains suffer in the following order: oats, barley, wheat, rye, maize. The airier and more siliceous the grain, the more sensitive it is to zinc damage.

The animal organism, too, finds zinc absolutely necessary. The organs rich in phosphatides and nucleoproteins are rich in zinc. Hair and milk are zinc-poor though the zinc content of milk rises considerably during the lactation period. Richer in zinc are muscles and bones. The liver is especially rich with 100 parts per million. Oysters, bony fishes, crabs, and starfish are rich in zinc. In the blood of the cuttlefish it is tied to protein, as is copper. Zinc has been found in cod-liver oil, in the muscles and liver of the sea lion and the sperm whale (about 40 ppm.). The herring, during its mating season, contains up to 160 ppm. in the body, and up to 350 in the testes. In swine and sheep the seminal vesicles are particularly rich in zinc, while in the bull it is the prostate, the sperm containing as much as 2000 ppm. It is striking that in snakes the poison glands show an increase in zinc. Presumably, this is to enhance the protein destroying effect of the snake poison, which is an attempt, as it were, to inaugurate an extra-bodily digestion by way of the bite. Strong decomposing forces are thus introduced into the protein. A decomposition of the protein, a segmentation or splitting of the protein of the nucleus preparatory to fertilization, is part of

147

the effect of the sperm; it drives this protein into chaos so that the creative, life-forming, cosmic, etheric forces may begin their work. It is in connection with these processes that we must look at zinc.

In human blood a zinc content of 0.5 to 2 ppm. has been found; in the organs five to six times as much as copper; in the liver of the adult, 15 to 93 ppm., in the brain, 5 to 16, and as much as 200 in the teeth. It is extraordinarily interesting that zinc is an important constituent of insulin, the secretion of the isles of Langerhans in the pancreas. Insulin is connected with the regulatory processes manifesting in the sugar content of the blood. In crystallized insulin we find 0.35 to 0.8% zinc. Too high an insulin content makes the blood poor in sugar, depriving the ego, the spiritual principle that can act in the blood by virtue of sugar, of the possibility of taking hold. This leads to coma. Again we refer to the fact that plants rich in zinc permit themselves no rich sugar metabolism, and that oats, which develop a starch that is particularly beneficial to diabetics (being especially accessible to the processes of light and silica), are far more zinc sensitive than any other grain. Zinc is also found in carbon dioxide anhydrase, and is thus allied to the process whereby carbon dioxide is lifted out of its combination with the fluid organization, devitalized, and made ready for exhalation.

In other important enzymes, such as peptidase, amidase, phosphatase, and phosphorylase, zinc has been found. With these substances as implements, the organism (which first created them as tools for its activity) carries out its work of decomposition. It directs the metabolism with the aid of such tools. The metals play a great role in this.

Taking zinc in an abnormal manner or in abnormal quantities can lead to striking physical, psychic, and spiritual phenomena. Among these are the occupational disease of the brass-foundry workers who inhale the metallic vapors, and the damage caused by continuous administration of small doses of zinc triturations. The barrier between the metabolic and the nerve-sense organization appears to be breached. The symptoms are congestions towards the head; flushing of the face; bursting of the small blood vessels in the same area, followed by prolonged bleeding; changes in the blood supply to the eyes; tendency to bleeding in the gums. Associated symptoms may be inflammatory processes around the eyes and ears; itching; swelling of the oral mucosa, the tongue, tonsils, and palate; coryza, running catarrh, pharyngeal catarrh, and salivation may follow. There may be inflammatory processes around the spinal nerves, with severe muscular pains. Disturbances in the body temperature, chills and fevers, indicate that the warmth organization is out of balance. Feelings of numbness and excessive drowsiness are experienced, but no real sleep will ensue, only a confused dozing. One feels exposed to two opposing forces, one of which would put us to sleep and the other awaken us. Our higher principles, the soul and spirit nature, are pressed out of the upper organization. We have encountered similar phenomena in connection with silver and copper. In the metabolic organization, the metallic process takes hold of activities carried on by the higher principles in unconscious depths, and displaces these upwards, setting them free. To this extent the zinc effects resemble those of the planetary metals "below the sun." They do not align themselves with regenerative processes, as copper does, for example, but with degenerative ones. Hence zinc stands alongside the planetary metals "above the sun." Decomposition is activated so that the higher principles may be released for their spiritual tasks to bring about consciousness and self-consciousness, to become "body-free." Zinc is perhaps especially suited for this purpose because it is a "mixed" metal, uniting within itself the impulses of the "upper" metals with those of the "lower" ones.

The growing child must gradually achieve the complete embodiment of his spiritual principles. His integration into life on earth starts with his physical body. Every seven years another of his principles becomes "ripe" for the earth until, with the twenty-first year, he reaches his majority and his spiritual personality is established. After his seventh year, his soul principle, the astral body, must prepare its entry. This may be accompanied by crises of development. The soul must battle for possession of the body. The result may be convulsive illnesses, ranging from whooping cough to chorea (St. Vitus Dance). Even the simple convulsions of dentition, epileptic cramps, and in the adult, asthma, spasmodic cough, and facial cramps, are all phenomena in which the astral body joins with the physical-etheric organization in an abnormal tensing, clenching way. These are fields in which zinc therapy has been tried.

XI

ALUMINUM

It is reported that Goethe, that tireless researcher, during the last days of his life asked to have a bowl of garden soil placed near his bed so that he might contemplate the secrets of that seemingly so simple but largely unexplored substance. Shortly before that a man had held in his hands for the first time the metal whose "ore" is the ordinary soil. Wöhler had just produced a minute quantity of aluminum by heating metallic sodium, a constituent of ordinary salt, with aluminum chloride. Around the middle of the century, in 1854, the young French chemist, Deville, exclaimed before the savants of the Academy of Sciences, "Just imagine how useful a white metal would be that was as stable as silver, did not turn black in the air, was fusible, malleable, ductile, and tough, and had the marvelous quality of being lighter than glass; of what great use such a metal would be if we could produce it in a simple way. When we consider, furthermore, that this metal is present in nature in the greatest abundance, that its ore is clay, then we must eagerly wish that it may become a metal of general use. I have every reason to hope that this will come about."

These prophetic words were vindicated. To begin with, further development led gradually to the use of aluminum for certain alloys, at a cost equal to that of silver, until Deville's hopes were fulfilled: "If we find ways and means of producing aluminum at small cost from its ore, which is the most widely distributed constituent of the earth's crust, it will become the commonest of all metals." This became possible when the invention of the dynamo brought up from the "unearthly" the electrical forces theretofore hidden in nature, and spread them throughout the world. Today we can say tersely that aluminum is clay plus electricity.

The laws of nature do not permit this unusual metal to appear in the pure state like the "classical" metals. One must apply enormous energy, all the splitting forces inherent in electricity in order to rob the natural aluminum ore (alumina or bauxite) of the oxygen with which it is so strongly connected and thus create the metallic condition. Gold, silver, quicksilver, platinum, copper, are present as "natural-born" metals; lead, tin, and zinc are easily smelted out of their ores, as though they were awaiting their liberation. But nature seems to expect aluminum to remain clay. The metallic condition is unnatural for it; the metal is not only difficult to extract, but the extraction would immediately be undone if a peculiar circumstance did not protect it from attack. Like an impenetrable armor, aluminum oxide immediately covers the metal with a fine protective layer of patina, a "noble rust." We may well call it this, for aluminum oxide as a mineral can achieve the noblest form of which the metal is capable; it can appear as corundum, sapphire, or ruby, those unusually hard and costly jewels. In contrast to the precious metals, the most precious condition of aluminum is not its purity, but its rust.

In nature, aluminum occurs primarily as clay, aluminum oxide, combined in manifold ways with other substances. Thus it has an important share in the formation of rock and of fertile soil. Without aluminum there would be no fertile earth. In the oldest igneous rocks it occurs as feldspar, and in feldspar itself it is a rhythmically mediating element between the two polar opposites of the siliceous elements and the calcareous (calcium, sodium, potassium). Feldspar again is an intermediate between the quartz nature and its opposite pole, mica, in the granites, and between the translucent quartz (which appears in crystallized form as rock crystals) and the light-reflecting shimmering mica.

Since about 60% of the earth's rock mass is feldspar, it is obvious that aluminum is not only the most widely distributed, but also quantitatively the most abundant metal. Based on 800 analyses of crystalline rock, the American geologist, Clarke, determined the aluminum content of the earth's crust to be about 8.0%, while iron was about 4.7%. Other estimates are somewhat lower, but all agree that aluminum is the commonest. Aluminum is also contained in certain varieties of mica (muscovite, biotite), and in certain hornblendes and pyroxenes (augites). The granites contain about 8% aluminum, the syenites 8–10%, and the trachytes about 8%, likewise the gabbros. In the micas, iron and magnesium appear and partly displace aluminum. From these rocks result, through weathering, the so-called secondary clay minerals. By miraculous life processes in which the lower animals play the chief role, these mineral earths are combined with the humus that is formed by low plant life from dead and decomposing organic matter. Thus arises the most important substance in the soil, the permanent humus, a semi-living thing in which all higher plant life can take root and prosper. What a marvelous instinct made Goethe reach for the substance in which is imbedded the secret of how nature has "invented death in order to have more abundant life." * Without clay, these processes would be unthinkable. It conveys to the soil the properties of plasticity and water absorption; the ability to unite with the living substance that is water, and to surrender to the plastic formative forces. Such is aluminum's role as a mediator. It is the instrument of a "mineral-plant" nature, as described in detail by Cloos. With its help the forces of the inner earth are carried upward into the plant world that unfolds upon the surface.

Let us once more return briefly to feldspar, the mineral in which the nature of aluminum originally manifests. It contains, on the acid side, silicic acid; as its basic constituent, alkalines and alkali earths, potassium, sodium, calcium; between these groups, aluminum oxide, clay, which can be either acid or alkaline (it can form acids with aluminum salts and alkalis with aluminates, and can itself become an acid). The metal of this igneous rock, feldspar, thus assumes a rhythmical middle position in it. Its activity and its properties make possible the "semi-life" of the "plant-mineral-nature" in the plant-bearing soil.

Let us recall that iron plays a rhythmical middle role in animal and man as a breathing metal in the ensouled world, and that the blood pigment exists, indeed, by grace of iron. A similarly decisive role, but this time in the plant world, is played by magnesium. Magnesium unites the light and airy with the liquid in the plant's process of assimilation. Iron unites the airy

---

* The reader is here referred to a book that gives a fresh view of geology as a science of the earth's process of becoming: Walter Cloos, *Life Stages of the Earth.*

and warm with the liquid and, simultaneously, the soul nature with the living body in animal and man. Aluminum unites the life-bearing water with the earthy element that is to be plastically formed. It is a bond between the silicious and the calcareous in the mineral world. Magnesium, in the green leaf, is a bond between earthly darkness and cosmic light, between root and flower; iron between ensouling and mere enlivening, between waking and sleeping, between upper and lower organization in the animal and human process. Thus aluminum is in a certain way the rhythmical metal of the mineral world, magnesium that of the plant world, and iron that of the worlds of animal and man.

It is with good reason that clay, uniting so readily with the life-bearing water, is the moulding stuff of the ceramic worker, the potter, and the plastic artist. The noblest vessels for holding liquids have from time immemorial been made of clay—majolica, faience, terracotta, porcelain. The sculptor works out his inspirations in clay before he hews them out of stone.

As a metal, aluminum shows properties that bring it closer to the metals "below the sun" than to those "above the sun." It is flexible, ductile, a good conductor for heat and electricity; it alloys easily with other metals. It is a gateway through which polarities are equalized in every way, be it those of heat and cold, of electricity, or of the metallic condition, standing as it does between the opposites of calcium and silica in the mineral world, or as an amphoteric element (in hydroxide) between acid and alkali in the chemical world.

In nature it follows, in its compounds, the ways of silica and oxygen, but not those of sulphur. Enormous energy is required to lift it out of combinations with silica and oxygen. One kilogram of the metal needs about 25 kilowatt hours. It generates intense heat in combustion, and liberates metals that are difficult to extract from their ores. In its salts it resembles iron in many respects, but it lacks the wonderful "breathing" relationship to oxygen, the play between the bi-valent and tri-valent conditions, for in its chemical behavior it is rigidly tri-valent. Alum, the double sulfate salt of aluminum and potassium, has the same property as copper sulfate of absorbing or ingesting heat radiations in its solutions, and thus releasing sunlight in a cooled state.

Aluminum's relation to silica, the form and light pole of the mineral world, is particularly apparent in the noblest form that aluminum can assume on earth. Here, like silica, it occurs as an oxide and attains not only the stature of a precious metal, but of a precious stone. The gem stones, with the exception of the diamond, are the offspring of silicic acid. But aluminum oxide is the only metallic oxide that can also become a gem stone as it does in corundum, in ruby, and in sapphire. The first is colorless, the last two are colored by metallic traces.

What is in the aluminum process that can kindle processes of healing? It is its rhythmical mediator nature, so clearly visible in the mineral world. Because of this rhythmical middle position, Rudolf Steiner indicated orthoclase feldspar (potassium aluminum silicate) and diaspore (a natural hydroxide of aluminum, hence clay-like) as remedies to support and strengthen a heart that has been weakened by illness.

We have not described from what regions aluminum was lifted so that in a few decades it could achieve its march of triumph through human affairs. The engineer can no longer do without this light metal, which can be given almost the tensile strength of steel and machined into essential parts for automobiles and airplanes. It has become indispensable to the housewife for easily heated, rustproof utensils of all kinds. Its good electrical conductivity, paired with its light weight, has enabled it to replace copper in high tension lines, and the forces that loose it from the womb of clay now flow along its veins and sinews all over the earth. It is the metal of the Electrical Age. The powers of the technocosmos incarnate in new, no longer nature-given materials. Aluminum was one of the first of these. The seven classical metals point in definite cosmic directions, manifest the forces of planetary beings, bring the cosmos to expression in the substances of the earth, as shown in the title of L. Kolisko's work, *Influence of the Stars on Earthly Substances.* Aluminum, however, as a "mixed king" put together from different cosmic effects in a way as yet unknown, has been, in respect to the cosmos, silent toward man up till now.

All the more has it been surmised from ancient times that the natural mother-substance of aluminum speaks to us in a spiritual language. When it assumed the form of a gem, it was worn, in those days, as a treasure of inner forces, a talisman and cultic adornment, and thus was precious to man. We are not speaking here of those traditions, no longer accessible to modern knowledge and consciousness, that have ascribed, half superstitiously, certain gems to certain months and zodiacal signs. The modern spiritual science of Rudolf Steiner directs us to the moral realms in order to build up a new relation to the precious stones. It points to the course of human destiny, to the will nature that works mysteriously and as yet unconsciously in man, creatively shaping the course of his life. The bodily organs that carry us toward our destiny are the feet. The sapphire is related to them. But when the will nature is no longer unconscious, when it is lifted into the full consciousness of knowledge, when it unites wholly with the essence of what is known—and for this we need unselfishness and love—then this will nature becomes the power of intuition. The will that has operated unconsciously in the flow of destiny therewith enters into the bright realm of knowledge, and at the same time into the region of freedom. The ruby is related to the force of intuition. The earth, made plastically impressionable by the clay within it, takes into itself the imprint of our footsteps.

If man acquires the spiritual forces that correspond to such noble natural substances, then a Moral Age will be added to the Technical Age that aluminum, as the conveyor of forces lying below nature (electricity and magnetism are indeed a world of such forces), must now serve. Such a Moral Age will then make use of natural substances with a consciousness of responsibility to the world.

XII

NICKEL AND COBALT—
SIBLINGS OF IRON

There are metals whose family resemblance to iron, in one or another aspect, is so distinct that we can speak of them in the same breath. Indeed, nature does just that when, for example, it combines iron, nickel, and cobalt in the meteorites. Chromium, manganese, the platinum metals—and more distantly tungsten, molybdenum, and titanium—likewise show many iron traits in their phenomena, so that we have strong reasons to suspect that the iron-Mars impulse cooperated in their formation. Even uranium may be included here.

In their atomic weights, nickel and cobalt range immediately alongside iron (iron 55.84; nickel 58.69; cobalt 58.94). Their classification in the periodic system is: iron 26, cobalt 27, nickel 28. In density, cobalt (7.7), iron (7.86), and nickel (8.5) are again close beside each other. The same sequence appears in their melting points; cobalt 1400° C., iron 1528° C., nickel 1542° C. All three are strongly paramagnetic, the most intensely magnetic of the metals, iron taking an easy first place, nickel second, and cobalt third. When heated (heat is a force of nature in polaric opposition to the electro-magnetic realm), nickel is the first to lose its magnetism, at 350° C., followed by iron at 768° C., while cobalt retains its magnetism until it reaches 1100° C. Cobalt and nickel, like iron, are hard and tough; nickel is more ductile, cobalt more tough. Like iron, these siblings can be forged and welded. Both are considerably more resistant than iron to atmospheric influences. In this they exhibit a feature that reaches its highest intensity in the platinum metals, making them almost precious metals. A cosmic imprint of a special kind is here added to the iron nature.

But although nickel and cobalt have ensconced themselves in the cosmic iron process manifest in the meteoric iron, they are not woven into the laws of earthly iron. The major iron deposits, above all the great iron ore belt around the north temperate zone, are not rich in nickel and cobalt. Some of the "greenstones," particularly those containing olivine, absorb as metallic infusions nickel, cobalt, chromium, and the platinum metals. Olivine is an iron-bearing magnesium-silicate, colored green by bi-valent iron. The olivines that are rich in iron contain more cobalt, while nickel predominates in the almost pure magnesium-olivines. On an average, olivine contains 0.2% nickel; but in the "acidic" primitive rocks (appearing with quartz and containing free silicic acid) only one-tenth of 1% is nickel. Nickel thus inclines more toward the magnesium process, cobalt more toward the processes of iron. Thus nickel has more affinity for the vegetative, and cobalt for the animal processes; for magnesium is the metal of the green chlorophyll, whereas iron is that of the red hemoglobin. Again we can find a key to this mysterious behavior in a cosmic process. Out of the cosmos come, in the meteor stones, not only iron but also mineral rock. This rock, however, is not like the ancient acidic rocks, granite, gneiss, or mica, but can be compared only to the "greenstone" formations, primarily olivine.

In his spiritual-scientific investigations of the evolution of the earth, Rudolf Steiner describes how in its "cosmic youth," when the earth was still much more intensely related to cosmic life and activity than it is today—being now a far more independent entity vis-à-vis the cosmos—the lifeless minerals were gradually precipitated out of the *life* of the earth. In his remarkable book, *Life Stages of the Earth*, Walter Cloos indicates that it was a stage especially permeated with the sun forces that laid the foundation for the "greenstone" formations in the earth. Three light-elements, iron, silica, and magnesium, joined in the formation of these minerals.

Garnierite, an important nickel mineral, is embedded in this iron-silica-magnesium sphere; it is a hydrated nickel magnesium silicate, $NiMg \cdot SiO_3 \cdot H_2O$. In it, nickel participates in all the olivine metamorphoses that appear as serpentine, asbestos, soapstone (steatite), and meerschaum. Garnierite has fine and tangled fibres; it is a kind of fibrous serpentine whose magnesium is replaced by nickel. Another metamorphosis is pimelite, a form of nickel soapstone or nickel meerschaum. Nickel is most receptive to the plant's fibrous nature, the plant-mineral condition found in these magnesium minerals. Another important siliceous mineral, chalcedony (a variety of agate), is able to offer an abode to nickel, and takes nickel into its own semi-precious stone existence, for chrysoprase, that beautiful apple-green semi-precious gem, is nickel chalcedony. The main deposits of these siliceous nickel ores, formed of weather olivine, are in New Caledonia.

These deposits, however, are far outranked by those in Canada, which are the largest in the world. Here, too, it is in the dark basic rocks (norite) adjacent to granite that the nickel ores—copper and platinum bearing nickeliferous pyrrhotites—have embedded themselves. Here, too, the iron siblings show their relationship to sulphur; it is magnetic pyrite and plain pyrite that harbor our two metals in the deposits in the Sudbury district of Ontario, Canada. Magnetic pyrite (pyrrhotite) carries more nickel, plain pyrite, more cobalt. Here we also find millerite (NiS); pentlandite ($FeNiS$); nickelous pyrrhotite ($NiFeCoAsS$); bravoite, an iron sulphide containing nickel ($FeNi)S_2$; but also braggite, a platinum-palladium-nickel sulphide. The relations of nickel to the platinum metals thus become apparent.

Cobalt occurs in other ways. Its most frequent ores are smaltite, a cobalt arsenide ($CoAs_2$); cobaltite or cobalt glance ($CoAsS$); and linnacite (Co_3S_4). Cobalt here combines with the arsenic process, much like iron, which forms arsenopyrite ($FeAsS$) as the most important arsenic ore, as well as loellingite ($FeAs2$). Nevertheless, the chief cobalt deposits are those that contain no iron! First among these are the African deposits, in the Congo and Northern Rhodesia; then come Canada and Burma. The African (and also the European) veins of cobalt-uranium-silver show this metal in close proximity to the silver and uranium processes.

Comparing nickel and cobalt with iron, we find that the relationships of iron to sulphur and arsenic appear distinctly in both these metals. On the other hand, these two siblings do not go along with iron in respect to oxygen, the carbon dioxide process, and the hydration process. There are no cobalt or nickel combinations corresponding to iron glance, hematite.

There is no cobalt or nickel spar. Neither has anything like limonite or Goethite. They lack what makes iron the breather among the metals and gives it a unique role in the processes of plant, animal, and man. They are onesidedly oriented towards sulphur and arsenic on the one hand, and towards silica on the other.

The chemistry of the nickel compounds shows the metal to be bi- and tri-valent, like iron. But nickel prefers the bi-valent state, for the nickel suboxide compounds are the stable ones. Nickel easily binds and transmits hydrogen, wherein it resembles the platinum metals. Because of this property it serves the chemical industry as a catalyst in the hydrogenation of fats. The nickel salts are mostly a beautiful green, but they turn blue with the addition of ammonia, and thus remind us somewhat of copper.

Cobalt, too, is both bi- and tri-valent in its compounds, and likewise prefers bi-valence. The bi-valent cobalt salts are a beautiful blue when anhydrous, but in the hydrous state they are peachblossom colored. The trivalent cobalt compounds are stable only as complex salts. They have a special relation to nitrogen and, with ammonia, form an extraordinarily large number of interesting cobalt ammines with yellow, orange-red, and violet-red colorings. Cobalt salts are added to the drying of linseed oil in order to oxidize this into varnish. Thus they transmit oxygen, in contrast to nickel.

The following blue cobalt pigments are important: smalt, an artificial fusion of silica, potash, and cobalt oxide; ceruleum, a product of cobalt oxide, silicic acid, and tin oxide; Thénard's blue, a cobalt aluminate.

* * *

Pursuing our twin metals into the realms of life, we find the following relationships, all discovered only recently. In plants there is an average of 0.01 to 2.0 mg.% of nickel and 0.02 to 0.13 mg.% of cobalt. Soils contain 0.4 mg.% of nickel, 0.1 mg.% of cobalt. In man and the higher animals, considerable nickel is stored in the liver, and the pancreas is relatively rich in it. Insulin is a thousand times richer in nickel than is the pancreatic secretion. Here nickel reaches into the immediate proximity of the zinc process. Marine molluscs are also said to be relatively rich in nickel. We find it, too, in the yolk of the hen's egg. All this points to the realms of the copper effects far more than to those of iron, to processes that take hold of the proteinaceous organization in its lower animal forms in order to prepare it for the iron process, for permeation by breath and heat, by the astral and the ego. Helpers and preparers of the iron process are what our two metals reveal themselves to be. Cobalt is present in the pancreas, but is especially abundant in the thymus gland and in the liver. It is the constituent metal in Vitamin B-12, which is important to life and indispensable to healthy blood formation. The constitution of this vitamin has only recently been clarified. It is an intensely red colored substance, closely resembling in structure the iron and copper breathing pigments, but containing cobalt as its focus instead of iron (in hematine) or copper (in hemocyanine). A lack of this cobalt substance leads to a serious blood deficiency, pernicious anemia, which was treated with liver extracts until it was discovered that the effective principle was the cobalt compound.

In polycythemia there is an overproduction of red blood corpuscles; the ego does not control the blood's economy or the iron process in the blood. Here cobalt can be a most important remedy. When the conversion of the protein in food into our own protein is disturbed (and this may be accompanied by serious dislocations of the entire protein metabolism), Rudolf Steiner recommended a novel remedy of meteoric iron in combination with pancreas. The natural combination of iron-nickel-cobalt is here brought into collaboration with the organ in which nickel and cobalt cooperate in a special way. (Zinc, in which the iron nature combines mysteriously with the planetary metals "below the sun," protects against the consciousness processes connected with degeneration, thereby making possible an undisturbed regenerative activity of the ego organization. Thus it works sedatively—"mineral opium.")

The higher principles deal with cobalt, nickel, and the other metals in an individual way. This is beautifully demonstrated by the metallic content of hair. Traces of Fe, Cu, Ni, have been discovered in all kinds of hair. In black hair, copper, iron, and nickel predominate; in brown and red hair, iron and molybdenum; in blond hair, titanium and nickel; in grey hair, nickel almost alone. It is also characteristic that the thymus gland, the organ of adolescence, the organ of the etheric body that must disappear when the astral makes its entrance by way of the thyroid gland and the sexual organs, is rich in cobalt. The soul principle living in the airy unites with the life principle active in the liquid, and iron enters into its birthright. Cobalt thus paves the way for the iron process, acting as viceroy in a "cuprous" phase of the organism's development.

A remarkable illness involving cobalt deficiency in cattle and sheep has appeared in Australia and New Zealand, the "pine" disease. Rapid emaciation is accompanied by a tremendous acceleration of the breath and pulse, failure of the mating instinct, and premature births. The astral organization (which brings about puberty) cannot properly fight its way through the breathing nature. Minimal quantities of cobalt provide a cure.

The cobalt and nickel processes thus clearly demonstrate in the human organization what they merely indicate in the inorganic. They act like an incompletely developed iron process. They have affinities to the sulphuric and arsenious, but not to oxygen and carbon dioxide; to the plastic-liquid protein synthesis, but not to the astralizing process working in the breath; to the gland activity, but not to the lungs and the nerve-sense processes. They can—indeed they must—prepare the organism for the iron process, clearing a path for it right from the start of the copper process, but they cannot replace it.

* * *

We must speak of one more thing with which cobalt is connected, the cyanogen process that is deeply hidden but important for the human organization. Here again we approach the processes of meteoric iron because cyanogen is found in the atmosphere of the comets, as Rudolf Steiner pointed out long before it was confirmed by spectroscopic analysis. The previously mentioned red cobalt color-substance, Vitamin B-12, contains a cyanogen group, in contradistinction to chlorophyll, hematine, and hemocyanine. Cyanogen, a carbon-nitrogen compound corresponding to carbon dioxide, is a terrible respiratory poison for all living beings. Nevertheless, it is a substance of fundamental importance in their physiology. Many plants, such as almonds and cherries, form and store cyanogen compounds in their seeds; many others form enzymes that split up these compounds (emulsins), thus pointing to an inner cyanogen process controlled so perfectly that all cyanogen, as soon as formed, is immediately reworked and none is stored. In human metabolism, too, a delicate cyanogen process is essential. Here also Rudolf Steiner was the first to direct our attention to this process, just as he was the first to point to the cyanogen in the comets.

By virtue of delicate cyanogen processes, forces unfold in the limb organization that resist the natural course of the other physiological chemical events, and call forth, as it were, a certain chaos, a purposeless space. Here the free will can find the bodily arena for its activity and imprint its intention upon the muscles, which would otherwise have to follow automatically the commands of the chemical energies dwelling within them. Rudolf Steiner gave repeated and detailed descriptions of how, by way of cyanogen—and the nitrogen laws of the ancient long-past stages of our earth that built up this cyanogen—the physiology of a being disposed for freedom is possible. In cobalt, iron sends its ambassador into the cyanogen sphere of the metabolism, so that this sphere may act in accordance with the blood processes.

*　*　*

The miners of former times, in certain conditions of atavistic clairvoyance, experienced in inner imaginations the elemental spirits of the world of minerals and metals; they had glimpses of the spirituality that forms these elements. They spoke of the elemental spirits of the solid and the liquid, etc., of gnomes, goblins, elfs, water sprites, and nymphs, some of which were friendly and some impish. To these miners the difference between gold and native arsenic was not one of densities, hardness, melting points, or colors, but a matter of beings who revealed their faces and features in these properties. In nickel they felt the connection with sprites and nymphs, in cobalt with the goblins. The "mischievous" nature of these beings was seen when ores that had looked promising volatilized with a poisonous stench as soon as they were smelted—which is what arsenic metals do with their arsenious constituents. Thus these ancient miners felt what it is that distinguishes our two metals from iron.

Until recently, therefore, these metals were largely ignored. Of nickel, we appreciated and used only its gem variety, chrysoprase; of cobalt, only its ability to produce beautiful blue glass. Now we have discovered their Mars aspects. We use the passive virtue of nickel for defensive weapons such as armor plate. We understand how to use cobalt plating to intensify the atom bomb as an offensive weapon of apocalyptic destructiveness. Cobalt's "destiny" of occurring together with radioactive uranium has been made even harsher, by extending the radioactive decomposition of the ura-

162

nium in the bomb, potentized, into cobalt itself. On the other hand, such artificially produced radioactive cobalt has now largely replaced radium in radiation therapy.

We enter a friendlier realm when we recall the healing virtues of our two metals, which are presented to us by the cosmos in the marvelous natural combination of meteoric iron.

XIII

ANTIMONY

A special earmark of this metal is its close connection with sulphur. Its most important ore, by far, is stibnite (Sb_2S_3), trisulphide of antimony. Of secondary importance are the four additional "sulfantimonites" that are the most significant among almost 100 antimony minerals. These are the dun-colored fahl ores or tetrahedrites, containing sulphur, copper, silver, and quicksilver, likewise iron, cobalt, nickel, and zinc; the red silver ore, pyrargyrite, a sulphur-antimony-silver compound; the lead-bearing jamesonite; and the iron-bearing berthierite. We should also mention two of some importance in mining: "Valentinite" and "senarmontite" (also called white antimony or antimony bloom). Both are SB_2O_3, the first crystallizing in rhombic and the second in cubic form. These are not primary ores, however, but products of the alteration of stibnite through atmospheric influences on surface deposits.

Antimony is inclined toward sulphur not only because it occurs mainly as a sulphide, but by its very nature. It is only half a metal, the other half being sulphurous. Its metallic aspects are underdeveloped compared to the metals heretofore described. Thus it is seldom found as a pure metal or in combination with other pure metals as in silver-antimony (dyscrasite) or nickel-antimony (breithauptite). To be sure, the metal is fairly dense (6.7), but it is easy to melt, easy to vaporize, even easy to burn. (Melting point is 630° C., boiling point 1440° C. in the open and only 735° C. in a vacuum.) A drop of the melted metal falling on parchment scatters quicksilver-like into droplets, which, while still burning, scorch all sorts of curves into the surface. The white oxide smoke of its combustion precipitates upon cool surfaces like hoar frost or ice flowers on a window. Combined with hydrogen, antimony forms a volatile and poisonous gas from which the metal, when liquefied by liquid air and the addition of ozone-bearing oxygen, can be derived in a state that may actually be called sulphur-like; it no longer looks metallic but is a bright sulphur yellow, weighs considerably less than in its ordinary metallic state, and is extremely active and unstable. In this process the resourceful chemist has, as it were, surprised it in the condition in which it once existed in volatile antimony-hydrogen. The contact with the normal conditions of today quickly transforms this sulphur-like state into the accustomed metallic form. An ancient condition, corresponding to previous phases of evolution, has fleetingly appeared.

Electric current is poorly conducted by antimony. Taking the conductivity of silver to be 100, that of antimony will be only 3.86. Its heat conductivity is likewise bad. But it shows strong forming forces; it has a foliated crystalline structure, related to the hexagonal crystal form. (The metals considered so far all crystallize regularly, tin alone being tetragonal.) Antimony is also brittle, and can thus be easily pulverized. Like quicksilver, it alloys readily with almost any metal, lending hardness and brittleness to the mixture.

Unlike iron, antimony is aloof to magnetism; placed between the two poles of a horseshoe magnet, it does not lie in a straight line between the two poles, but diagonally. It is diamagnetic, in contrast to such paramagnetic metals as iron, nickel, and cobalt. But it is not only passive toward magnetism; it also rejects electrical forces in a curious way. When electrolytically refined from a solution of chloride of antimony, it precipitates on the cathode as a metal in the form of a blackish powder. When scratched, rubbed, or heated, this powder changes with "thunder and lightning," i.e., with radiations of heat and light and small explosive noises, back into the normal antimony form. This explosive antimony is less formed and also lighter (density 5.8) than the normal metal. It has retained certain forces of heat, light, and levity by which it defends itself against the gravity of the world of matter as well as against the sub-material world of electricity. Only in connection with tin have we encountered modifications of such pronounced character. Tin responds to intense and prolonged cooling by changing into the *light,* grey tin-pest. This must be viewed as resistance to the cold processes. The explosive defense against electricity on the part of electrolytically produced antimony must be viewed in the same way, and this is how Rudolf Steiner interpreted this strange behavior.

Antimony again behaves strangely when it is cooled close to the freezing point and is just about to change from the liquid state into the solid. It expands on solidification, becoming lighter rather than denser, and thus indicates a resistance to the firm condition of the earth. Here it resembles water, which contracts in cooling down to 4° C., but then expands and becomes lighter. Thus ice is lighter than water and floats in it. Antimony alloys, when poured into a mould, fill the tiniest details. Therefore they are used in the casting of type metal, which is an alloy of antimony and lead. Thus antimony, like water, defends itself against becoming "earthy." (Earth is here meant as what constitutes solidity in the ancients' theory of the elements.) Like water, it desires to remain connected with the cosmos to retain the more cosmic liquid state, which it loses in the process of rigidification to a larger degree than do the flexible inwardly "liquid" metals. It becomes inflexible, "coy," towards the firm, purely earthy condition, and in solidifying it also takes on the same hexagonal crystal form as ice. This reveals another relationship to the etheric-liquid realm of the forces raying in from the cosmos. We do not find the formative force of the cube, but that of the six-pointed star, the snowflake.

These inraying forces come to expression most clearly in the chief natural ore-form of antimony, stibnite. Its long thin crystal needles look like materialized lines of force, bundles or interwoven formations of rays. They are like a metallic manifestation of the rays of the newer projective (synthetic) geometry, which weave the most diverse geometrical forms out of infinity. Antimony, "strives, wherever it can, to assume a tuft-like formation. It thus forms itself into lines that strain away from the earth and towards the forces working in the etheric. By the use of antimony we introduce into the human organism something that meets halfway the workings of the etheric body." In antimony "there is a tendency to enter the etheric element the moment conditions for it are ever so slightly present. It easily adjusts to certain force-streams in the earth's surroundings." *

---

* Steiner-Wegman, *Fundamentals of Therapy,* Chapters 16 and 20.

Stibnite seems entirely metallic with its density (4.6) and its brilliant metallic lustre. It melts easily, even in the flame of a candle (at 550° C.), so that it can easily be melted out of its ground mass or gangue by the iron reduction process. The fine-fibred crude antimony is obtained in this way. Stibnite also vaporizes early, assuming a "smoke" form, out of which it precipitates on smooth glass tubes in a lustrous condition. It ignites readily and gives off the heavy white smoke of antimony trioxide.

Summarizing all of the above, we might call antimony "sulphur that has become metallic." It feels completely at home in the realm of sulphur, and appears most happily as antimony trisulphide. Even the pure metal is half metallic, half sulphurous. No wonder that it produces no imposing phenomena in the realm of the salts. The alchemists of the declining Middle Ages, who occupied themselves extensively with the mysteries of antimony, regarded salt, mercury, and sulphur as the three forms of earthly substantiality. In the salt condition they saw above all the forces of the earth and in the sulphuric, the forces of the cosmos. Mercury was the connecting principle, weaving back and forth between the two opposites. Antimony was felt to be strongly present in the sulphurous, less so in the mercurial; while in the realm of salt it lost its power.

* * *

Antimony ores are widespread, but we must take them all together as a totality, as the earth's "antimony body" (Rudolf Steiner). Then certain laws will emerge. The continents are far from being equally rich in antimony.

The most abundant and purest ore, predominantly stibnite, is found in East Asia, particularly China (Hunan Province). The Japanese stibnite crystals, sometimes attaining the length of several yards, are well known. During the last century Asia has produced 46% of the world's supply of antimony.

Next come North and South America, especially Mexico and Bolivia. Their mines have yielded about 25% of the world's antimony in the last hundred years. They contain stibnite and many of the fahl ores.

Europe, producing about 22% in the same period, takes third place. The principal countries of origin are France, Hungary, and Austria. In the mines of Schlaining (Burgenland in Austria), Europe can boast of the oldest important antimony workings in the world.

In contrast, the African and Australian deposits, with their yields of 4% and 3% respectively, have been insignificant.

* * *

The Far East, the extreme West, and the European central region are thus the three principal members of the antimony body of the earth. Its center of gravity lies unmistakably in East Asia. In the Americas the metal appears in the regions that are richest in silver (and in lead), as described in our chapter on silver. Europe, the middle zone of this antimony organism, is the quicksilver continent. Thus antimony also dips into the quicksilver sphere. Indeed, antimony and quicksilver occur together in Italy (Monte Amiata) and Russia (Nikitovka). The American, Asiatic, and Australian deposits are near the great copper belt around the Pacific Ocean, with which we have dealt in our chapter on copper. Thus the antimony organism reaches into the regions of silver, copper, and quicksilver.

The complex sulfantimonites, second in importance only to stibnite as sources of antimony, all point to silver, copper, and quicksilver. This is especially true of the fahl ores, which contain all these three metals. (As to the gold content of some antimony deposits, more may be read in our chapter on gold.)

Geologists have noted that antimony occurs most often where old primitive mountain formations have been intruded by younger ones and tilted fractured zones have resulted. This is the case along the Andes cordillera, the folded zones of the Himalayas, and the younger European mountain formations. The antimony deposits seem to be relatively young Tertiary formations. Where solidifications have been gripped anew by strong dynamic forces, where a certain rejuvenation has thus taken place, stiff form and the will to change have impinged upon each other. This makes the earth receptive to antimony.

* * *

The data we have accumulated on antimony can now be illuminated by Rudolf Steiner's spiritual-scientific research. We spoke about this briefly in Chapter 2, but a more detailed presentation is now in order.

Investigation of the cosmic aspects of the metals results in the correlation of the seven principal metals to the seven planets, as described at the beginning of this book. The lesser metals are due to the collaboration of two or more planets. For one of these metals, antimony, Rudolf Steiner indicated the specific planets whose influences meet in it. In *Spiritual Science and Medicine*, we find in Chapter 19 the following:

> What works in antimony is really present everywhere. The anti-monizing force, if I may coin this expression, is at work everywhere. This force acts in man in a regulatory way, but so that normally he draws it from the outer-earthly. He draws, as it were, from beyond the earth what antimony concentrates within itself. Since in his normal condition man does not turn to the antimonizing force on earth, to what is concentrated in antimony, but to its outer extra-terrestrial force, the question naturally arises: what is this antimonizing force beyond the earth?
>
> This force, speaking from the planetary point of view, is the cooperation of Mercury, Venus, and the Moon. When these planets work together rather than alone, their effects are neither mercurial, nor coppery, nor silvery; they work as antimony works on earth. To establish this we must investigate the effects on man of aspects in which the three forces of Moon, Mercury, and Venus neutralize each other by being either in opposition or in square to one another. This mutual neutralizing brings about the same interaction . . . that the earth makes use of in antimony. For in everything that antimony is on earth there works, from out of the earth, the same force that these three planetary bodies exert upon the earth from outside.

* * *

This discovery throws light on the above phenomena. The "antimony body" of the earth is oriented towards silver, copper, and quicksilver. A mineral such as tetrahedrite or fahl ore, the pale sulphide containing copper, silver, and quicksilver, becomes, as it were, the interpreter for the cosmic forces that were once active in forming it. It is more difficult to find the mercurial, the silvery, and the coppery in the properties of antimony itself. The "metallities" of these substances have been neutralized by their inner "oppositions and quadratures." But antimony's volatility, its great versatility in forming alloys, its ability to scatter and disperse, all show something mercurial. Something silvery shines in the antimony lustre. If the sum of the atomic weights of these three metals (copper 63.57, silver 107.88, and quicksilver 200.61) is divided by three, we obtain 124.0, which comes closer to the atomic weight of antimony (121.76) than to that of any other metal.

*　*　*

Anthroposophical research yields a picture of evolution in which the beings of the various kingdoms of nature no longer arise and develop unrelated to one another. The evolutionary phases of man, in his soul and spirit as well as in his body, tally with the phases of these kingdoms. From the beginning, there resounds in man's evolution the leitmotif from which the melodies of animal, plant, and mineral result. The creative forms of the extra-human kingdoms are extracted and separated out of the human creative process. Man, the first-born of creation, sloughs off many forces and formative impulses. He cannot, for example, retain within himself the plethora of life as of the first day; he must sacrifice a great share of it (out of which the plant kingdom is formed) or he could not develop the consciousness that makes him a self-conscious spiritual being. Thus certain forces had to be ejected from his nature into the development of the kingdoms of nature, forces that would have been incompatible with the development of *the conscious human will*. These are the unconscious forces that work so strongly to form and shape the metabolic-limb organization, which enables the human will to work. These superabundant forces, banished into the mineral kingdom, have led to the formation of antimony with its strong plasticity. In *Spiritual Science and Medicine* Rudolf Steiner explains:

> You see . . . in the mineral kingdom we have something in antimony that has an inner kinship with the human will, in that this will, the more conscious it grows, feels more and more compelled to bring forth a countereffect against the effect of antimony. . . . everything that works organizingly in man under the influence of the thought forces, . . . notably the unconscious thought forces . . . as they still work unconsciously in the child, . . . is sustained by the forces of antimony; with all this antimony cooperates.

Antimony stimulates the forces of the inner organs, that is, those shaping forces which, out of the formless proteinaceous process, mould the forms of the organs. In the human blood we can see especially clearly the struggle between the form-dissolving will forces and the formative "unconscious thought-forces." The blood must remain liquid, formless, but it must always be ready to submit to the shaping forces of the organs in the organism, which it is obligated to nourish and build up. This expresses itself in its clotting ability. In hemophilia this ability is lacking. Bleeders have too little antimony force in their blood.

Therefore antimony is an important remedy wherever the forming forces have become too weak, wherever there is deformation in the organs, wherever the body's protein tends too strongly towards the dissolving and too little towards the formative forces. "By virtue of the properties we have described, antimony can convey to the blood the form-engendering forces of the human organism." In the healthy person the astral body transmits moulding forces similar to those of antimony, centrifugally from within outward—as manifested by the blood-clotting faculty. It opposes the centripetal forces in the fluidity of the blood that are encouraged by protein. If the "antimonizing" tendencies of the astral body are too frail as against the proteinizing ones, antimony will have a healing effect. If the activity of the astral body is too feeble to sustain the brain and sense processes—which can show up as dulling of consciousness and even as somnolence—antimony will lend support and cause the return of memory and initiative. "The organism is regenerated by the strengthened soul" (Steiner-Wegman, *Fundamentals of Therapy*).

The attentive reader may suspect a contradiction in antimony being described first as a complex of forces of the Mercury-Venus-Moon impulses, and then as an effect of the play of shaping and dissolving forces within the human organism. The contradiction will dissolve if we recall what was said in the copper, quicksilver, and silver chapters on the synthesizing metabolism, the formation of protein, and the preparation of blood being just the fields in which the impulses of these three metals can intervene. It is only when they have been rendered mutually "opposing" or "neutralizing" that the intrusion of the shaping formative forces becomes possible; the antimony effect builds upon the effects of silver, copper, and quicksilver.

171

The History of Antimony.

Antimony entered into culture a long time ago. Vessels are known to have been cast from it in the time of the Sumerians. These may have been used, like the "eternal pill" of the Middle Ages (a small antimony pellet), for curative purposes. Liquids stored in them for a long time would absorb traces of the metal, which would supposedly then produce powerful effects upon stomach and bowels. Copper was alloyed with antimony for hardening purposes. Finely powdered antimony served as a cosmetic to darken the eyelids, not so much for outer beautification as to bring the power of sight in contact with the antimony-force, for the use of these ointments was restricted to priests and kings. In the Middle Ages antimony, like quicksilver, became the metal of physicians and alchemists. Gold was refined to the highest degree with its aid. Basilius Valentinus (16th century), in his *Antimony's Triumphal Chariot*, described many areas of application and the preparation of many compounds. Paracelsus knew how to use it as a great remedy. The alchemists regarded it not only as an accessory in purifying the precious metals, but as a substance in which a special meeting of nature and of man could be experienced, a meeting called the creation of the homunculus. The meaning of this has mostly been misunderstood. Rudolf Steiner has explained that what happened here was a clairvoyant experience of the forces of antimony together with those of one's own proteinaceous organization. "There appeared to them, (the alchemistic physicians) in the process that they carried out in the laboratory when antimony unfolded its forces, projected into it out of their own nature, something that fights against the antimony forces as proteinaceous forces" *(Spiritual Science and Medicine)*. Thereby they experienced the human form in the dynamics of the metal.

XIV

THE NATURE OF SULPHUR

The way of observing the world of matter that is in vogue today describes the single substances according to their attributes: density, color, form, etc. This way may be called a static, even death-like approach, for it describes only what a substance is, by and for itself. For the future, we need a dynamic approach that reckons with life and provides an understanding of what a substance means in the realm of life, and above all in the human body.

In the present book on the nature of twelve metals, their behavior toward sulphur was a guiding thread by which the observation could ascend from the static to the dynamic and from the mineral realm to the world of life. Therefore, it seems proper to devote a closing chapter to sulphur.

We must grasp not only the telling properties that sulphur possesses, but also those that it does not possess—if such a paradox may be permitted. Rigid selfishness keeps to itself. What is to be of service to another must be approachable, must become an instrument, must lay aside self-existence and be capable of transformation.

In the self-centered direction, sulphur has its density (2), its hardness (2.5), its crystal form (rhombic), its melting point (114° C.), its boiling point (444° C.), its yellow color, and its atomic weight (32). But it can easily change, or even abandon, most of these properties. It is sensitive to its surroundings, especially to heat, but also to mechanical and chemical influences.

At 95.6° C., for example, it strips off its compact rhombic constitution and assumes a prismatic monoclinic form. At 114° C., it altogether relinquishes its solidity, turning into a thin yellow liquid (the so-called gamma sulphur) that yields to many solvents and turns brown on further heatings. At 160° C., it changes into still another type, mu-sulphur, which is viscous and dark brown. At 400° C., it becomes a thin liquid again, and when poured into cold water, it solidifies to plastic amorphous sulphur, which is not soluble in carbon bisulphide as are the previous modifications. At 444.5° C., it boils and turns into a bright gas. Thus it is easily affected by heat and strongly modified by it in the formative forces; it yields to heat, and at the same time defends itself by becoming, as a liquid, denser, darker, and tougher the more it is heated.

Sulphur reacts with equal sensitivity to the chemical nature of its surroundings. It combines with alacrity, burns very easily, and oxidizes slowly even at normal temperatures (as shown by its sour odor, particularly when pulverized). Combined with hydrogen, it becomes a gas; with carbon, it becomes a volatile ether-like liquid; with nitrogen, it turns into a volatile explosive mass; with the halogens, chlorine, bromine, and fluorine, it forms volatile liquids, even gases. But its enchanted light-nature shows best in its manifold combinations with the metals and in the characteristic colorings that sulphur imparts to them. The metal sulphides, taken collectively, constitute a complete rainbow of colors.

Opened to its surroundings in this way, closely akin to heat and light, withdrawing so readily from the solid and mineral, reminding us more of resins and pollens than of minerals, sulphur appears to be much more in tune with the forces raying in from the cosmic surroundings, the universal forces, than with those raying out from the center of the earth. Not without reason did the ancients call it sulfur (akin to sol, the sun), or even theion (connected with divine forces). When we are closely acquainted with the nature of sulphur, it is startling that it should be in solid form, for it ought to be a kind of air. We must see it as a paradox, as "solid gas" or "frozen hot air."

This is due to a second aspect of its nature, one that would like to contract the environment into itself and make it into a center, a focus. For sulphur inclines in a high degree towards self-densification, self-combining, toward "polymerization."

174

Sulphur actually should be an oxygen-like gas. It has the same ability to combine with everything; it is related, like oxygen, to what is most contrasting; it takes hold of the acid-formers as well as of the alkali-formers. It can do this because it is polaric within itself, hence can offer to every outer polarity the inner and opposite one. But, in contrast to oxygen, sulphur likes to saturate, to neutralize the polarities within itself, to combine with itself. Compared to the "selfless" oxygen, it is an "egotistical" self-enjoying substance. This ability to unite with itself makes it possible for fiery sulphur vapor to condense as a solid body. (On the basis of these observations, and numerous others that need not be entered into here, the chemists have given to solid sulphur the symbol S_8, pointing to an eightfold combination with itself.) Further tendencies toward polymerizations are shown in the many thio-hydrogens, polythionic acids, etc.

Sulphur is omnipresent. It exists in the air (every piece of tarnished silver betrays this); in the sea (as sulphur salt); in the numerous sulphur springs; in spring and river water (as sulphate of lime, though only in traces); as sulphates (of potassium, sodium, calcium) in the salt beds. Calcium seizes and "calcifies" it, even causing it to appear as a rock in gypsum, or to become semiprecious in alabaster or selenite. But these minerals are much too soft to maintain their condition. The anhydrite hills and gypsum mountains formed in some areas never attain to the serene solidity of the ancient rocks; a sponge-like swelling, a slow dissolution grips them, a kind of rotting, by which the sulphur is released. This results in the formation of sulphur springs and of enormous sulphur deposits. The sulphur process manifests even more impressively in volcanic activity, in which it enters the air again. Thus we see that the forces of solidification, the world of the mineral, are able only in a transitory and incomplete way to fetter this mobile flighty substance. The metals have imposed on it the most rigid chains, binding it to their mercurial spheres in the form of pyrites, glances, or sulphides, but no sooner does such an ore enter the light of day than there are endless transformations. Air and water take hold, dissolve, absorb it into their more transient spheres; vitrioles are formed; lime solidifies it for a time, but sulphur wrests itself free again, appears as an element, disappears again, oscillating continually between surrender and self-assertion; with the help of the volatilizing forces of hydrogen it rises into the air, only to be thrust down again by the oxygen that burns it.

175

Through this capacity of entering into the play of the forces of the surrounding world, of straining to meet the inraying sphere, and at the same time contracting, densifying, summoning up the earthly forces of matter, the outraying central forces—by virtue of these dual properties, sulphur can play a significant role in the world of life. All prototypal protein, the primordial substance of life, contains sulphur. Neither liquid nor solid, swaying between form and formlessness, incessantly seized and released again by the life forces, regenerating itself and again being broken down by the life processes, protein cannot do without sulphur. Protein is continually revived and continually dies again. It is taken hold of by the forces of the organic totality and, because of its lack of any strong traits of its own, because of its all-around formability, is the ideal material for the formative forces. Then, however, it is claimed again by the earthly forces of matter and brought near to the lifeless. The forces of the cosmic circumference, the "formative force world," synthesize the living protein; the forces stamped upon matter by the earth complex analyze and break it down. With this interplay of polarities, sulphur has a profound connection by virtue of the properties we have described. It promotes the enlivening process of protein. Two of the important amino acids arising in even the most delicate decomposition of protein—cysteine and cystine—are sulphur bearing.

Sulphur and Plant Life

The nature of sulphur comes to expression most clearly in two plant families, the Cruciferae and the Liliaceae, especially in the subgroup of the onion family, the Alliaceae. In these plants we find an enormous vitality, a proliferating plasticity that almost bursts its bounds and is gorged with juice, but nevertheless refuses to submit to a well-defined form or rigid shape. This surging and swelling, however, eventually shoots into vigorous blossoming processes and dissolves into color and air, often into a sulphurous odor. The stinging nettle deals differently with sulphur, displacing it into the inorganic down in the roots. Therefore, nettle protein decomposes easily and rots quickly. Yarrow, on the other hand, achieves an "exemplary" combination of sulphur with the potassium process, and

176

camomile does the same with calcium. In equisetum there is a special collaboration between the sulphur and silica processes.

Sulphur is also present in the protein of animals, particularly in the areas of the peripheral silicic acid processes such as in hair, horns, claws, in fact, in every horny formation. But this is the utmost solidification attained by sulphur. It avoids any mineral hardening such as, for example, phosphorus achieves in conjunction with lime in forming the bones. We find sulphur only in cartilage as chondroitin-sulphuric acid (amino sugar, oxidized glucose, acetic acid, ethylsulphuric acid). Ethylsulphuric acid is also found in urine, and it binds the poisons produced by decomposing protein in the intestines (indole, phenol, cresol).

Sulphur is contained in Vitamin B_1, which is present in starchy fruit and yeast, and is of such vital importance in controlling the carbohydrate process and combating beriberi. This vitamin, together with phosphorylase, works in animal and man as an instrument for proper carbohydrate combustion.

The human organism uses sulphur to detoxify the cyanic process by forming non-poisonous thiocyanogen (cyanogen sulphide) compounds.

In the human organism sulphur is active throughout the entire body, from the digestion up to the nerve-sense organization, in a widely ramified way. In the digestive process it aids the absorption of extraneous protein (wherefore we generally season protein-rich foods with plants that contain sulphur, such as onions, chives, garlic, radishes, horseradish, mustard). Sulphur then promotes the enlivening of what has thus been assimilated. It makes all physical processes susceptible to the formative force organization (the etheric body). Too much sulphur over-stimulates the etheric life activity, to the detriment of the consciousness-engendering processes of the ego and soul. Too much mere vitality results, and too little consciousness. The drowsiness found in the spas with sulphur springs impairs the consciousness.

To the now vivified protein, however, sulphur gives an impulse toward aeration; it assists the connection between blood circulation and breathing and helps to carry what is airy into the liquid realm. It follows the tracks of protein, but goes much further, reaching right to the periphery of the organism where, in the skin, nails, and hair, it stagnates and becomes

177

horny. The various metals carry the sulphur effects toward the regions of the body that belong to them, just as in the great organism of the earth they have tied down sulphur in their deposits.

Rudolf Steiner has described the importance of sulphur in the proteinaceous sphere with especially characteristic words. Sulphur "trans-homeopathizes" protein; the spiritual building forces (and beings) use sulphur as a sculptor uses water to wet his fingers before kneading the moist clay into form.

The principle of homeopathizing rests upon the overcoming of the material nature of the substance to be homeopathized, upon releasing the dynamic processes frozen into matter. The forces of cohesion, which are the outraying earth-center forces, are to be annulled, and the forces raying in from the circumference, which are closely affiliated to the substance to be potentized, are to be drawn down and tied into the potentizing medium. The substance is changed from a three-dimensional to a two-dimensional condition by the constantly enlarging surfaces brought about by the shaking, rubbing, etc. The rhythm dwelling in the substance is set free. The substance acquires more and more "skin," more planes of contact with the surrounding world, and thereby it grows sensitive to the inraying (etheric) world of forces.

Basically, all these processes are akin to those that vivify protein. Sulphur is the substance that, by virtue of all its properties, enhances these vivifying homeopathizing processes. Thus the spirit can use it to "moisten its fingers," in order more easily to put its mark upon the earthly material.

It is interesting to note that patients in whom a too dense physicality impedes the efficacy of the delicate homeopathic remedies, are treated by homeopathic physicians initially with sulphur, in order to open the way for the force complex that is "akin to a spiritual nature." (Hahnemann viewed the nature of the homeopathic preparations in this way.)

Sulphur thus brings the substances into function, unchains their dynamics, as soon as it goes to work in the medium of the living protein. In doing so it clings mainly to the volatile hydrogen nature; it refrains from combining with oxygen to form acids and salts, which would drive the living substance into a mineral state as is done, for example, by phosphorus.

It is no accident that all the elements present in air are also present in the decomposition of proteins: oxygen, nitrogen, carbon (in the form of carbonic acid), hydrogen (as water vapor and, especially in high altitudes, as free hydrogen), sulphur (as traces of hydrogen sulphide or sulphurous acid). Rudolf Steiner described the early stages of the earth, the "infancy" of our planet, as a much more life-filled, much more rarefied existence, pulsed through by spiritual forces from the surrounding spheres. (To imagine these conditions we must contemplate the living kingdoms of nature, rather than the mineral kingdom in which nature has already died.) In these surrounding spheres there was a strong plant life, weaving atmospherically, cloudlike, in a delicate "proteinaceous air," a substance that was denser than today's air, but far finer than today's fluids. This proteinaceous atmosphere was pervaded by intensive sulphur processes; in this superabundant life, so closely connected with the cosmos, sulphur could give free play to the forces of its essential nature. To such conditions the names sul-fur or theion fit well indeed.

But if the created earth was to become the habitation of self-conscious spiritual beings, i.e.,'of mankind, the overpowering life forces had to be stemmed and death processes had to be injected into the earth's evolution. The inraying planetary forces sent their metal-forming etheric streams toward the earth; these combined with the sulphur in the proteinaceous atmosphere and precipitated it toward the solid earth. The intervention of the death processes brought about mineralization, solidification, and density. The formation of the oldest rocks occurred simultaneously.*

The properties of sulphur hint at these creative events of the earth's beginning. The primordial relationships of sulphur to protein and the metals continue to exist, like runes of the first days. Sulphur is an "atmospheric" element, but its ancient atmosphere died, decomposed, and in its decomposition became today's atmosphere of oxygen, nitrogen, carbon, hydrogen, sulphur. But whenever these dead elements come together to form living protein, sulphur is again in its true element. We must look at it "vitalistically" and not mineralogically. Only then will it reveal its true secret.

---

* For fuller descriptions see Rudolf Steiner, *An Outline of Occult Science*; G. Wachsmuth, *Cosmos, Earth, and Man*; W. Cloos, *Life Stages of the Earth*.

FOR FURTHER READING

Rudolf Steiner intended these carefully written volumes to serve as a foundation to all of the later, more advanced anthroposophical writings and lecture courses.

THE PHILOSOPHY OF SPIRITUAL ACTIVITY by Rudolf Steiner. "Is human action free?" asks Steiner in his most important philosophical work. By first addressing the nature of knowledge, Steiner cuts across the ancient debate of real or illusory human freedom. A painstaking examination of human experience as a polarity of percepts and concepts shows that only in thinking does one escape the compulsion of natural law. Steiner's argument arrives at the recognition of the self-sustaining, universal reality of thinking that embraces both subjective and objective validity. Free acts can be performed out of love for a "moral intuition" grasped ever anew by a living thinking activity. Steiner scrutinizes numerous world-views and philosophical positions and finally indicates the relevance of his conclusion to human relations and life's ultimate questions. (262 pp)   Paper, $8.95 #1074; Cloth, $18.00 #1073

KNOWLEDGE OF THE HIGHER WORLDS AND ITS ATTAINMENT by Rudolf Steiner. Rudolf Steiner's fundamental work on the path to higher knowledge explains in detail the exercises and disciplines a student must pursue in order to attain a wakeful experience of supersensible realities. The stages of Preparation, Enlightenment, and Initiation are described, as is the transformation of dream-life and the meeting with the Guardian of the Threshold. Moral exercises for developing each of the spiritual lotus petal organs ("chakras") are given in accordance with the rule of taking three steps in moral development for each step into spiritual knowledge. The path described here is a safe one which will not interfere with the student's ability to lead a normal outer life. (272 pp)   Paper, $6.95 #80; Cloth, $14.00 #363

THEOSOPHY, AN INTRODUCTION TO THE SUPERSENSIBLE KNOWLEDGE OF THE WORLD AND THE DESTINATION OF MAN by Rudolf Steiner. In this work Steiner carefully explains many of the basic concepts and terminologies of anthroposophy. The book begins with a sensitive description of the primordial trichotomy: body, soul, and spirit, elaborating the various higher members of the human constitution. A discussion of reincarnation and karma follows. The next and longest chapter presents, in a vast panorama, the seven regions of the soul world, the seven regions of the land of spirits, and the soul's journey after death through these worlds. A brief discussion of the path to higher knowledge follows. (195 pp)   Paper, $6.95 #155; Cloth, $9.95 #154

AN OUTLINE OF OCCULT SCIENCE by Rudolf Steiner. This work begins with a thorough discussion and definition of the term "occult" science. A description of the supersensible nature of the human being follows, along with a discussion of dreams, sleep, death, life between death and rebirth, and reincarnation. In the fourth chapter evolution is described from the perspective of initiation science. The fifth chapter characterizes the training a student must undertake as a preparation for initiation. The sixth and seventh chapters consider the future evolution of the world and more detailed observations regarding supersensible realities. (388 pp)   Paper, $9.95 #113, Cloth, $16.00 #112

To order send a check including $2 for postage and handling (NY State residence please add local sales tax) to Anthroposophic Press, Bells Pond, Star Route, Hudson, NY 12534. Credit Card phone orders are accepted (518) 851-2054. (Prices subject to change)

A catalogue of over 300 titles is free on request.